Living Spirit's

Guidebook for Spiritual Growth

A Program for Spiritual Transformation

Jef & Tanya Bartow

New Paradigm Publishing

Copyright © 2005 by Jef Bartow

New Paradigm Publishing, llc
P.O. Box 589, Bayfield, CO 81122
970 - 769-4919

npp@livingspiritcommunity.net

Cover Design by Tanya Bartow
Cover picture by Jenny McCarthy

All rights reserved.

Printed in Canada.

For information about permission to reproduce selections from this book, write to New Paradigm Publishing, llc, P.O. Box 589, Bayfield, CO 81122.

Library of Congress Cataloging-in-Publication Data

Bartow, Jef & Bartow, Tanya
Guidebook for Spiritual Growth/Jef & Tanya Bartow
p.   cm.

Includes bibliographical references.

ISBN 978-0-9760863-1-4

1. Spirituality   2. Self-help   3. New Age Non-fiction

First Printing

This book is printed on acid-free paper.

Table of Contents

Introduction..................................1-8

Joy..9-20

Being I Am...............................21-32

Appreciation............................33-47

Trust.......................................48-61

Prosperity...............................62-78

Faith.......................................79-91

Mindful Loving.......................92-109

Job's Journey.......................110-119

**Changing Altitude through
Changing Attitudes**................120-132

**Compassionately
Helping Others**......................133-147

Decision-Making..................148-159

Filled with the Spirit............160-170

Epilogue..................................171-172

Suggested Readings.................173-174

Introduction

Spiritual growth in all its various venues is about leading us into more and more refined and powerful spiritual energies. To prepare us for these non-ordinary and intense psychological states, we need to simultaneously purify our Personality as a part of our spiritualizing process. Leading an accelerated spiritual growth life beyond human evolution is the spirituality taught over thousands of years in both Eastern and Western mysticism, and metaphysics.

A powerful way to bring our spiritual nature into the outer material world comes through our efforts in developing and expressing Love, Wisdom and Will. Simply, Love is abundant generosity; our capacity for relatedness and a coherent attractive giving that makes all things whole. This Love part of Spirit expresses differently at various levels within our human universe. By living Love where we are, it will lead us to receiving and expressing all the various levels of Love.

Wisdom is internalized within ourselves through the integration of knowledge with experience. By reading, studying and learning, we develop knowledge. By applying this knowledge, we develop Wisdom. Without Wisdom, our life journey will most probably lead us to a dead (unspiritual) end. With the integration of knowledge and experience, we become a sage in our life, and for others.

Finally, developing Will within our material nature (Personality) is accomplished through both discipline and sacrifice. Executing spiritual exercises which develop will helps us transform our personal will into spiritual Will. We must ultimately give up our personal will in order to live God's Will. By living God's Will for our life, we become co-creators in fulfilling God's creativity for our planet and life within it.

This Guidebook

There are many books and literature from the past and present which can help us learn about spiritual growth and spirituality, whether they use these terms or not. They can be books on philosophy, Western theology, Eastern mysticism, psychology, metaphysics, and self-help to name a few. The important point is that we can find all kinds of valuable resources to help us develop knowledge about spiritual growth.

This is why you will **not** find this guidebook to provide you an extensive or comprehensive knowledge base. An appendix of suggested readings is included to help you along this program, as needed. The intent of this Guidebook is to **provide you practical**

spiritualizing exercises which will turn your knowledge into Wisdom. The design of many of these exercises will also help you transform your personal will into spiritual Will.

The most powerful use of this Guidebook for spiritual growth is to continually approach each spiritual practice with Love. This is love of yourself and those you will encounter. It is giving and relating with generosity in order to make your self a unified whole person. It is also lovingly helping those in your life to be and express their own spiritual nature. A key spiritual principle which will maximize your spiritual growth in this program is to:

Always keep a balance of how you live the process with getting to the end results.

Specifically, both success and failure are integral parts of spiritual growth. Without failure, we will never get to the end result. Spiritual growth in Matter is challenging and challenge will always bring instances of failure. On the other hand, without success we sooner or later lack the motivation to continue. This is why we need to embrace both our successes and failures. Another spiritual principle applies here: **we can learn and grow far more from our failures than we will from our successes**. So, besides maintaining an orientation of Love throughout this program, we need to embrace an orientation which accepts failure as well as success along the way.

By living this program for spiritual growth with love and non-judgment, we will accelerate our growth and Wisdom. This does not take away from the reality that we do need to fully execute each spiritual practice.

Giving up or skipping ahead when the going gets tough only affirms the human orientation that life and growth should be easy and/or comfortable. Frankly, changing our outer material orientation to a spiritual one is never easy or comfortable. In the end, it is intensely satisfying and fulfilling. This fulfillment includes joy, love and a variety of experiences communing with Spirit and ultimately our God Within.

For many, this Guidebook will seem like jumping into a river without a boat or paddle. If you have not already participated in The Quest Program or the Course in Miracles, this spiritual growth program may seem very challenging. There is no requirement that you have completed these prior to beginning this *Living Spirit* program. On the other hand, during this program you need to feel that it is both challenging **and** achievable. If it is not challenging, then there will be no sense of achievement. If you cannot feel that it is achievable, you will most probably not finish the program.

Spiritualizing Energies

One final characteristic of this program which differs greatly from many others is the energy by which this Guidebook has been developed. Consistent with everything within the *Living Spirit*™ Series is the spiritual energies available for you on your journey. As you feel these energies, they will tend to make everything seem more difficult. When you embrace this Spirit, you will find a whole new level of insights and illumination provided to help you in your spiritual transformation. This is why it is a very good idea to re-read and reflect on each exercise and practice along the way.

At the beginning of each spiritual practice is a quote, principle or truism. It is intended to set the tone for this set of exercises. Using this principle as part of your daily meditation or reflection will help you align with the energy available for this spiritual practice. As you might expect, living the process of this Guidebook is about immersing yourself into each spiritual practice until completed.

Regarding our opening principles, they are "the origin or cause of something; a fundamental truth or motivating force upon which others are based." A principle also includes an essential element or quality which produces a specific effect. In our case, the principle includes an energy which will produce a specific spiritualizing effect.

A further way to help you embrace the Spirit of each spiritual practice will be a definition of the keyword or title. A definition is "a statement of what a thing is" and "a putting or being in clear, sharp outline."

In our Guidebook, the principle for each spiritual practice will provide a fundamental truth that will open up the spiritual energy of the practice. The definitions in each practice will help bring a clear and sharp outline to what this spiritual practice is. The intent is not to bring you an intellectual comprehension, but to help you align with and embrace the energy of the practice. So remember, energy first, comprehension later.

The prelude explanation to each spiritual practice is to help each of us use our psychological facilities of sensing, feeling, thinking and intuiting to fully embrace the energy available for each practice. In a sense, this explanation is to help make the principle and definitions more warm and fuzzy. As mentioned, the suggested readings will be a good way to augment this short and concise prelude to each spiritual practice. Therefore, if you feel somewhat agitated by the lack of substance here, remember that it is the not the intent of this Guidebook to provide you knowledge, but to facilitate your gaining Wisdom.

Each of the 12 spiritual practices will involve a number of exercises. Each is labeled a Step. Simply, this is a step-by-step spiritual practice to facilitate transformation. Your Wisdom and transformation will come by executing the steps. Another key principle here is: **if you must understand the end result at the beginning, there will be no growth involved**.

So, let yourself feel like you're on an adventure into the unknown of your spiritual growth. It is not how correctly you execute each step, but the attitude, orientation and effort you make along the way. The energy will lead you and provide opportunities for illumination. And if Spirit behaves as usual, your insights will not come when you want them to or in the way you expect them to.

A powerful practice in spiritual growth is to document your insights and illumination along the way. The process of formulating and documenting your insights helps internalize the spiritual energy involved. It also helps reinforce the Wisdom gained. It is strongly recommended that you document insights on a daily and weekly basis. Just like dreams that are not documented quickly, a little time and normal daily life drastically affects the clarity and comprehension of your insights.

So, either keep this Guidebook, a journal or tape recorder nearby throughout this program.

As in school, many times there are opportunities to put forth additional effort to gain additional insight and learning. For some of our spiritual practices in this Guidebook, there will be the opportunity for extra Wisdom.

Here, the effort required will be more strenuous and the opportunity for spiritual accomplishment more intense. A fairly easy way to put it is that these exercises will not be for the faint of heart or will. They will be a good signpost of things to come on the Spiritual Path.

Some final notes, each spiritual practice is structured to provide a sense of accomplishment, either during or by the end. If you do not feel a sense of accomplishment, then something is incomplete. In this case, it is a good time to get feedback from your mentor, facilitator, or teacher. If that is not practical, then use someone that you trust as a sounding board. With the right orientation, the energy will always lead you to the truth of how to bring a positive sense of accomplishment with each spiritual practice.

Each spiritual practice is also designed to be accomplished in about a one to two month period of time. This does not include your reading or outside study as a learning process. A good wake-up call that there is significant interference, resistance or opposition to your progress will be if you're significantly behind this ballpark time frame.

If so, it's a good time to reflect and meditate on your internal obstacles to moving forward. Having done this, then look to become aware of external influences that are affecting your progress. Making changes to eliminate these obstacles and influences is a spiritual practice in itself. Frankly, the Spiritual Path is fraught with such obstacles, opposition and influences. Learning to lovingly deal with them is a positive way will powerfully accelerate your spiritual growth.

Joy

> *This is the day the Lord has made. Let us rejoice and be glad in it.*
>
> *Psalms 118:24*

As our first spiritual practice, I believe it is safe to say that all of us have experienced joy in our life. It is fairly easy to describe this joy as gladness, happiness, glee or excitement. Joy gives us a sense of pleasure. For most of us, experiencing this joy is a routine part of daily life.

So why have joy as a spiritual practice? Normally, joy is a reaction to an experience. It is a vicarious or purposeful result of some activity. We experience joy as part of a birthday celebration. We laugh at a TV comedy. We read something that tickles our funny bone. This emotional based experience is a fundamental part of many day-to-day activities. Joy as a spiritual experience goes far beyond this.

Spiritually, Joy is a very powerful spiritual state which we can penetrate, and thereby express regardless of our activity or experience. Eastern mystics have a term for it: Ananda. It is ecstatic joy, higher than the highest human joy. It is the "delight of self-existence." Joy is the flame of inspiration, the "bliss energy of God." This divine Bliss produces "absolute freedom without a binding term or limit."

Western mystics also describe a mystic joy or "all-surpassing joy," a "joy of illumination" which brings the "attainment of That Which Is." Our normal consciousness is inadequate to describe this heavenly joy. It is a state of "utter peace and wild delight." "At this level the soul swims in the sea of joy, that is in the sea of delight." This joy is also the "stream of divine influences."

To get an idea of this spiritual joy, try to remember specific experiences when you laughed so hard you almost cried. Remember a type of hysteric state that would set in and seem like it could go on and on. At these times, I remember that when I could finally settle down, I was exhausted, as though I expended a tremendous amount of energy. Indeed, this highest human joy is an energetic state, just like spiritual joy.

To contact the state of spiritual joy, we must transcend our human experience of joy to enter a state of exaltation, rapture, ecstasy, exhilaration and jubilation. We have heard mystics write about it, but the experience is otherworldly. For many, this self-bliss is the highest spiritual experience attainable in life. This exalted state is described by Sri Aurobindo as:

> "This Delight, which is so
> imperative, is the delight in the Divine for
> his own sake and for nothing else, and for
> no cause or gain whatever beyond itself."

Our spiritual exercises for this practice are intended to lead us up to and into these spiritual energies. In doing so, we get a taste of living within Spirit while still functioning in the outer world. As you progress, hopefully you will experience a joy which just spontaneously wells up within you from time to time.

Tanya's Wisdom

It has always been easy for me to have joyful experiences. This changed in my mid-thirties. Until then I had buried the abusive parts of my childhood, keeping them in the closet of my sub-conscious. The more I resisted letting the memories come to the surface, the more impossible life became to endure.

Finally, the barriers broke and I was flooded with one horrific memory after another. As I worked to heal those parts of me, I dropped into a black pit. It was suggested that I take an anti-depressant to help soften the difficulty. A side note to this suggestion was that it would also slow the healing process, so I refused. I wasn't going to numb what had been numb far too long. I looked for other ways to shift me from my malaise.

One day while standing in the kitchen washing dishes and trying not to wallow in the depths of my emotions, a blue bird flew to a nearby ledge and sang it's proclamation to life. I could not help but laugh and feel the joy beginning to well up within me. What a perfect gift!! Such a simple thing, but it brought with it a transformation of my mood. This feeling of joy was different from what I had experienced in the past. It seemed to be generated from within. I can't say I was able to hold to this, but it did give me hope.

Later that same season as I hit another particularly bumpy spot, a hummingbird flew up to my sunglasses and stared me eye-to-eye. He was defying me not to sink into the darkness, but to sing joyously at the possibilities that lie ahead. I took that spunky little bird's challenge. It seems from that point on there have been so many joyous treasures to be found that all I have to do is look.

Step One: See joy in every day.

As just mentioned, this spiritual practice is about working up to a spiritual peak experience. Consistent with much of spiritual growth, we must begin where we are. So, for the next week or so, focus your attention each day on examples of joy.

At work, you might see it in the break room while your fellow employees are taking a break from the daily work schedule. If you have children, it should be fairly easy to see them laughing joyously while watching TV or reading a book, or just playing. You may find it at the local coffee shop while you're waiting for your coffee. Look for it in the smiles of those that are serving you, whether at the grocery store, your local retailer or restaurant.

If you can, document it as you go through the day in your journal or tape recorder. If not, use your daily reflection time to record these experiences and your awareness of them. You might even set a goal for this week to see an increasingly number of joyous experiences each day.

For yourself, use at least one day to do something which follows from an insight from Osho: "Play means doing something for its own sake." So, bring some play into your life and experience it with joy for its own sake.

Insights: *(From your past week, write your conclusions here about joy in life around you)*_____

_____
_____
_____
_____
_____
_____
_____
_____
_____

_____
_____
_____.

Step Two: Experience joy in everyday moments.

      Here we build on the joys from our first week. Now that we can see joy in daily life, let's experience it through others and life around us. Life is a series of small accomplishments, whether large or small, whether significant to others were not. How often have we seen a sense of joyous glee at that moment of accomplishment? So, let's move beyond our observance of joy to experiences of joy.

      Watch for these small accomplishments around you and experience the joy involved. To begin, this could be your children's small step forward, or a coworkers finishing up a project, or that service person's reaction to a sincere thank you. Feel joy with them and for them. Tap into their joy. See how many times a day you can feel joy just based on the joy around you.

      And we're not done yet for this week's exercise. Now that you can feel joy in everyday moments, whether yours are others, let's be proactive about joy. Start with your home life. Bring joy into your environment and those who live with you through some small activity or celebration. If you live alone, change something about your home to express more joy. If you live with a partner or family, engineer some delightful experience for all without them knowing about it beforehand.

Watch and experience their joy as they spontaneously react to your efforts. Beyond this, do something for your coworkers out of the joy in your heart. It can be as simple as bringing cookies or treats for break time or, even more heartwarming, by you baking something you know they will enjoy. And even more importantly, experience joy as you do it. Feel how you would feel if they did it for you. Immerse yourself in joy.

Insights: *(How did joy expand for you this week? How easily did it become a part of your experiences? If not easily, why are you not yet fully open to joy)*_____

_____
_____
_____
_____
_____
_____
_____
_____
_____
_____
_____
_____
_____
_____
_____
_____
_____
_____.

Step three: Bring joy into your spiritual practices.

It's now time to focus on one of the four Noblest Qualities of Mind from Eastern mysticism. It is to live a subtle state of being joyful and rejoicing. For most of us, many of our spiritual practices require looking deeply within us, making changes and facing uncomfortable truths. Not necessarily joyous experiences. This may seem like an oxymoron, but living our spiritual practices can be a most exhilarating joyous part of our life. So let's learn how.

First, let's realize that our spiritual practices are time and experiences with God. In a sense, it is our communion with God. So make it a communion with joy. By making our practices filled with joy, we will also make it about love. Joy will lead us to love, to loving our spiritual practices each day.

Now, list those spiritual practices that you do at least on a weekly basis. Good examples would be journaling, reading, mindfulness or meditation. Others could be exercising, your diet regiment or how you interact with others. Some practices we will introduce during this program are watching movies and TV, socializing and even living our hobbies.

How do we go about making our spiritual practices joyful? For many of us, a sense of accomplishment is our goal for our spiritual practices. For this next week, make it about experiencing joy as you do your spiritual practice. If you're like most, it will take some effort.

If you have trouble in the beginning, then just recall your experiences of joy during this overall spiritual practice. When you're documenting your insights, experience the joy of your insight and what it means for you. If you're on a particular diet as part of your spiritual practices, savor each bite and experience the joy of being full and satisfied by this food. Even easier, maintain a state of joy while you're taking time for your hobbies. And if you can, translate your efforts in exercising into a joyful experience; get the endorphins flowing and try to reach a state of self-delight or peaceful bliss.

Insights: *(What obstacles in you tried to block your joy? What external influences seemed to get in the way? What breakthroughs did you have?)*_____

_____
_____
_____
_____
_____
_____
_____
_____
_____
_____
_____
_____
_____
_____
_____
_____.

Step Four: See joy in all of life.

      We all have a choice on how we see life. It is one of the fewest choices we have on the Spiritual Path. For some, life is about the strenuous efforts, struggles, failures and the fleeting experiences of happiness along the way to death. For others, life is idealistically about being in love, happiness and the rewards of family and making a contribution to humanity. From these two contradictions, mystics and spiritualized humans convey both the strenuous efforts, difficulties and failures as well as the love and bliss that we can live and experience on our journey back to God.

      During this last exercise of our spiritual practice of Joy, it is time to turn our sporadic experiences of joy into living joy consciously day-to-day. This will help us make Joy an internal part of how we live life. Spiritual Joy is part of how Spirit *dances* with Matter. When we see joy in all of life, we see Spirit in our material universe.

      So, wake up with joy! It is our choice to wake up with joy or no joy, love or no love, freedom or attachment. Make it Joy this week. Let go of your issues, conflicts, the manure of life. Our world, our life, our dreams, our joy are all of God. Look for joy in each moment. Put on your searchlight for joy. Look for where it is hiding. It doesn't take a miracle, only an orientation.

      For the next week, take the time to see and feel joy

in nature. Look past outer appearances to see how plants enjoy their life. See the joy in how they reach to the sun for nourishment. See the joy in their expression of beauty as they flower and procreate. Give them joy and watch and feel their response.

Even easier, express joy to the animals in your life. Watch how they delight in your attention and connection with them. Expand your joy to them and experience their delight. Do this until you can appreciate and immerse your self in their delight and the delight of being alive.

For humans, see how joy is always under the surface, ready to jump out. See those who are waiting for joy in weekends off, upcoming vacations, a dessert or anticipated dinner. Watch how moments of despair and sadness can turn to joy with the right empathetic comment. Watch for a mother's smile or a father's glow. Experience how your love given without requirement triggers joyful delight in those around you.

Do this every day and all day long until you can be Joy. Sing it to the mountains, the rivers, the sun and moon. See a *dancing* universe of joyous experience. As you make the efforts to reach this ecstatic state, anticipate the "Ah ha" experience. And when it hits you, go with the flow. Run the trail of your joyous experiences until a spontaneous explosion of joy hits you like a powerful river. Let go of your limits on Joy.

Insights: *(Where did you most feel joy this week? What did you learn that shocked you the most? How did you see Spirit dance with nature and us humans?)* _____
_____
_____

_____
_____
_____
_____
_____
_____
_____
_____
_____
_____
_____
_____
_____
_____
_____.

Extra Credit: None is needed if you got to an "Ah ha." If you haven't yet, then keep on seeing joy in all of life until you do.

Being I AM

I am a point of light within a greater Light.
I am a strand of loving energy within the stream of Love divine.
I am a point of sacrificial Fire, focused within the fiery Will of God.
And thus I stand...
    Affirmation of a Disciple

This first stanza of an ancient mantra is a way for each of us to assert our soul identity as part of a focused group of disciples in a new age. Each of us is a point of light as part of a greater Light. Each of us can express a part of divine Love. Each of us can be part of God's fiery Will in service to humanity.

How we do this depends on the uniqueness of our Soul's qualities and character. The more we become that uniqueness, the greater contribution we can make. Over numerous embodiments we have been developing personalities to express our unique soul qualities. Through these soul experiences, we have been making our Soul a self-conscious point of light.

Today, in this embodiment, each of us has chosen to express a part of our total Soul character. This part of our Soul in embodiment is what is traditionally called our individuality. It goes beyond the conditioning of our childhood and our building of a Personality. The easiest way to contact this part of our total self is to go deep within our heart. By doing so, we can come to the place of our uniqueness.

Astrologically, this individuality is represented by our Sun energy. Of the 12 parts of our Soul, we choose a dominant energy (Sun sign traits) to qualify our individuality for this embodiment. As we live this energy in daily life, we more and more express this part of our Soul. This expression within our Personality brings us part of our needed Wisdom for this life.

Psychologically, our individuality is defined as "the principle which makes possible, if need be compels, a progressive differentiation from the collective psyche." Our individuality motivates us to become an expression of our Soul's character. As we detach from the collective, we consciously incorporate more of our Soul into who we are day-to-day. On the Spiritual Path, we transform our entire psychological and material nature to become our Soul's character in full expression.

This spiritual practice is designed to help us identify and live our Soul's character. Through self-examination and release, we can open the door to more fully being who we are as a Being.

Tanya's Wisdom

I have had many roles in life. Going from child to mother to where I am today. Along the way I've rebelled against what society said I should be, this was those teen years. In junior high, I wore a tie to school before Annie Hall made it a fashion statement. During this phase, I was looking to shock people. In the south, it really didn't take much. On the other end of the spectrum, there was a big part of my life where I molded myself to accommodate the life I was living.

As I began my spiritual journey, I began to look at myself to define "who am I." I was no longer the maiden bursting with a thirst for life and adventure that I had exuded in my youth. Since my divorce, I was no longer the trophy wife who needed to be at my best to stand out and help my husband. I began to look for me. Who was I? This wasn't an easy thing to come to and it took a lot of trial and error.

I looked at my material possessions. More specifically, at my jewelry as a means of reflecting who I was. I now found wearing gold repulsive, since it reflected the monetary side of the life I had lived. So I abandoned all the gold I had accumulated over the years and bought inexpensive jewelry. That was easy.

As I began to furnish my new home, I decided to

start with a clean slate. I saw this as an excellent chance to let go of the traditional style furnishings I had used during my marriage. I wanted someone to walk into my home and get a real feel of the true me. I focused on items that were eclectic, unique and feminine. This was a lot of fun. I explored funky shops for little treasures that were different and unique. In my youth I was looking for a shock factor. Now I was looking for things that reflected my uniqueness.

    I then turned my attention to my wardrobe. I went to my closet and got rid of a third of my clothes; which wasn't that difficult. Maybe that's because what I let go of were things I had held on to since the 80's. And I really wasn't into those huge shoulder pads any longer. After saying good-bye to the 80's, I went shopping for that new me.

    My first attempt was to try the earth mama look using some organic clothing lines. These clothes didn't express me; I looked frumpy. I also let my hair go to its natural grey. So, I not only just looked frumpy and twice my size, I looked 10 years older than I was. That was going from one extreme to the other, which hit my ego.

    With this failure under my belt, I decided a better approach would be to look for words that defined who I was. Words that seemed to work were confident and refined, yet casual. I began to shop with these words in mind. Sometimes my past habits got in the way and I bought clothes that didn't match my key words. Overtime, I learned that these would usually sit in the closet, because I didn't feel comfortable in clothes that didn't reflect me.

Having moved away from the earth mama look and with some concrete words to use to express myself, I once again faced my closet. This time it was more difficult. I took every piece of clothing and asked some very hard questions. How long had I had it? If I had worn it 10 years ago, wasn't it time to let it go? If I hadn't worn it in a year, it also needed to go. Did I really need 12 evening gowns? Which pieces reflect my key words; confident, refined yet casual? This time I weeded out another half of my clothes.

I still morn the loss of some of those awesome clothes. Were they me? No, but they had been great friends. Today my clothes could use another level of weeding. But the way I dress does reflect who I am. I'm no longer putting on a façade of who I should be.

I've also come to realize that the key words I used in the past related to expressing the image of who I am. The actual qualities of my I Am are joyous, loving strength. I now see these are the inner characters that brought me through many of the difficulties in my life.

Step One: Review your life through the eyes of others.

For the next week, use your meditation time for a mini-life review. As you begin, separate yourself from you. Try to become your parents, teachers, friends and

25

coworkers and see how they see you.

Begin with your childhood and how your parents, teachers and friends saw you. How were you different, not conforming to other kids? How were you alike, normal let's say, or a follower? What stood out about you? Were there any key decisions about you that jump out? What changes affected you most?

Then move to your adolescence to how you changed with puberty and more life experience. How were you different and the same as your friends during this intense period? What impacted you the most about leaving childhood and becoming a young adult?

Next, view yourself as an adult and include the new people in your life, especially teachers and coworkers. Finally, see yourself today through their eyes as you have matured and progressed in life.

You need not be highly concerned about whether you're using terms that are roles you played or traits you expressed. It is important that you try to see which terms for you have lasted from early on and which has changed over time. For some, it will be as though you're a flower beginning as a bud and moving toward full bloom. For others, there may be powerful traits you exhibited as a young child that have mellowed over time. And there are always cases where in order to get along, you had to be who they wanted.

Insights: (*What character traits have you exhibited since childhood? Which have become stronger and stronger overtime? What qualities of yours have met the most resistance in life?*)_____

_____
_____
_____
_____
_____
_____
_____
_____
_____
_____
_____
_____
_____
_____.

Step Two: Confirm or revise your view based on feedback.

Here, interview those involved in your vision. Ask them to give you three traits that best describe you during each part of your life. Let them describe you to you. Be open and inquisitive. Laughingly elicit non-flattering and/or negative traits or behaviors. Ask them who they would have liked you to have been. Tell them that you want to honestly look at your past.

Additionally, try to find an old friend or classmate or teacher you can interview. Also include an old girlfriend or boyfriend if possible. The intent here is to look and see a clearer you from outside.

Look for what you didn't know. Look not to confirm, but round out and fill in; what surprised you! Take time to reflect on your erroneous vision of yourself. See if you can identify why part of your view was off the mark. Get a vision of you that is accurate, truthful and honest.

As you bring a clearer picture into mind of yourself, realize that this is your "personal center." It is the features that make you who you are. From this clearer picture, now document a description of yourself including traits, qualities and roles. Include traits that linger from the past and those you're trying to cultivate. In the end, your description needs to be an honest picture of yourself, as if you were documenting your epitaph.

Self Description:_____
_____
_____
_____
_____
_____
_____
_____
_____
_____
_____
_____
_____
_____
_____
_____
_____.

Step Three: Mold a positive only image of who you can be.

Now it's time to visualize our positive individuality infused with our Soul's character. Central to this picture are your most positive qualities and traits in life. Look for consistencies in described qualities and traits, especially those different from your parents and conditioning. Key in on traits and qualities, not the roles you've played. How did you relate to others, especially the opposite sex.

Explore your positive character traits and how they would express on a daily basis in each area of your life. Explore your passion of who you are and how it would express without limitations. Be sure to include those traits that you know you have, but did not express normally.

"A man's character is his guardian divinity... Character is destiny."
Heraclitus

This character is your "self-possessed power." It is who you are in a state of "tranquil eternal repose." It is your real nature underlying all your material expressions. It is your "inner man, innermost reality and central connection to God."

When completed, use your meditation time to

release your past. Release your past conditioning that still affects you presently. See those limiting factors in life evaporating. Feel yourself release the emotions which tend to produce your most negative behaviors. Work on this visualization until you can clearly see your most positive character.

From this positive real you, now write out a description of this you. This is your I Am.

I Am:_____
_____
_____
_____
_____
_____
_____
_____
_____
_____
_____
_____
_____
_____
_____
_____.

Step Four: Step out and live this I Am who you are.

"A man's true estate of power and

riches is to be himself; not in his dwelling, or position or external relations, but in his own essential character."
Henry Ward Beecher

Begin this last Step with your three most powerful character traits. What can you do each day to express these? Is there a new activity that will best help you live these traits? Who can you rely on to help support you being who you are?

Next, take a look throughout your home environment to see how it does or does not express these three traits. What changes can you make to help your living environment better reflect you? Now look your wardrobe. What part of it expresses these qualities and what part reflects your conditioned past? Finally, does your automobile reflect this you?

If you're not in a position to make some major changes here, then make some small changes. Change your bedroom first to better express this inner you. See how you can change your workspace to reflect your positive traits. In addition, post pictures on your refrigerator of (home, car, clothes, etc.) that which will better express the real you at sometime in the future.

Look for obvious ways you don't live these qualities in current work, family, social and romantic situations. Then change them! And all the while, hold this positive picture of you somewhere in the back of your mind and feelings. You are this I Am. What you don't express today, you will in the future.

Insights: (*Write down the first three changes you will make immediately to be your I Am. Next, right down the three big*

changes you will make as soon as possible.) _____
_____
_____
_____
_____
_____
_____
_____
_____
_____
_____
_____
_____.

Extra Credit: Identify three of your personality traits or current ways you relate to work, family, friends and a partner which do not fit with your I Am. Then, change how you interact with others, behave and live today to eliminate these conditioned, or lingering parts of your past.

3 Personal Traits & Changes:_____
_____
_____
_____
_____
_____
_____
_____
_____
_____
_____.

Appreciation

"Appreciation is a potent message and invaluable tool for all of us, not in the old sense of 'be grateful for what you have,' but in the new awareness of how life is a wondrous opportunity."
       The Power of Appreciation

      For so much of life these days, we are subtly conditioned to take both what we have and the freedoms and opportunities provided for granted. For the majority of us who have not had to survive a depression or world war, it's fairly easy to make assumptions about what life will bring or give to us. As the pace of objective materialism increases, it's also hard to slowdown and savor the moment.

As the above quote demonstrates, appreciation goes far beyond being thankful for what we have. If you follow the teachings in *The Power of Appreciation*, it becomes a way of life, a total attitude and orientation to life. Appreciation is a way to transform your self and all aspects of your life. Through the attractive nature of appreciation, you will also attract to yourself what you desire and cherish.

Simple definitions for appreciation are "to think well of; to be fully and sensitively aware of;" and "to recognize and be grateful for." Within Eastern mysticism, the first of five "mindfulness trainings" on the Noble Eightfold Path is "reverence for life." This is valuing and appreciating all life. It is a grateful recognition of the part that every living being and thing provides in God's creativity.

I have come to conclude that appreciation is one of our highest emotional experiences within our material universe. Spiritually, we have the opportunity for even greater mystical experiences. But related to nature and our day-to-day life, feeling appreciation and proactively expressing it relates to our developing awareness, recognition and caring for our world around us.

Incidentally, I have come to define our various senses (25 Personality senses) as those faculties, powers, and capacities for becoming aware of and transforming our environment. Therefore, it's only a short step to make appreciation one of these senses in which we become able to see the aesthetic beauty in all of nature and life.

Since appreciation is a power or faculty, we have the potential to develop it like we do our five physical senses. Just as we learn to savor, enjoy and care for various foods based on taste, our sense of appreciation is our "attitude of savoring, enjoying, appreciating, caring, in a non-interfering, non-intruding, non-controlling way" for all of life.

In the beginning, we usually appreciate what we value and are grateful for. As we develop our sense of appreciation, we can learn to appreciate everything. As Abraham Maslow put it: the end product of aesthetic perceiving (appreciation) is that everything is "equally savored, and in which evaluations of more important or less important tend to be given up." Unfortunately for many,

> *"Our conscious intellect is too exclusively analytical, rational, numerical, atomistic, conceptual and so it misses a great deal of reality, especially within ourselves."*
> The Farther Reaches of Human Nature

So, for the next month or so, let's explore the energy and feeling of appreciation which can lead us to further developing our sense of appreciation.

Tanya's Wisdom

In the fall of 2000, I moved my daughters to a large town and began to study for my Masters of Fine Art. Life was insane. My father had been diagnosed with lung cancer in the spring. That summer I traveled

800 miles every 2 weeks to care for him. On the weeks I was home, I bought a house and hired a crew to do renovations while I was away. During this time I also squeezed in packing and the move. My father died right before school started. I had little time to settle in.

After that, life didn't calm down much. Each day consisted of more than 2 hours of commuting, taking my daughters to and from school. Being a single mom, there was always homework to make sure was complete and dinners to make. The MFA program also took an inordinate amount of my time. My stress level was over the top.

I began to let parts of my life slide. I still had not found time to settle in after the move. I had rooms stacked with boxes. My spiritual life suffered the most. I just did not have time to concentrate on what I claimed was <u>the</u> most important aspect of my life. I had stopped reading anything of a spiritual nature. I no longer attended workshops and I also stopped meditating. The only thing I did that related to a spiritual activity was attending an occasional yoga class.

My life was in chaos when September 11, 2001 happened. Like most of the country, I was stunned. I didn't know anyone personally involved, but that didn't matter. I've never reacted strongly to these types of things. But this was different and I didn't know why. I was devastated. This was a wake-up call.

I had come into this life to accomplish some things and I was stumbling around as if I had all the time in the world. I thought that I would get back to my spiritual purpose when time allowed. But with 9-11, *death tapped me on the shoulder and said, "I can take you any time, any place."*

I was overwhelmed with appreciation for the sacrifice those individuals made. Thousands had laid down their life to tell me (and the world) to wake up. I gathered with friends to say thank you for those individuals' sacrifice. These gatherings weren't to mourn, but to recognize their contribution.

I began to change my current situation. I had to get with the program and do it now. I ended a relationship that was a dead end. I began yoga teacher certification. I began to take stock in all areas of my life. What would their sacrifice have meant if I went back to my life, pre 9-11?

Within 2 months, I met Jef. Here was someone who truly lives for spirit. Had 9-11 not happened, would I have been ready for that life-changing meeting? Probably not.

Today, life is still intense, but I now have a greater appreciation for the little things. I have gone from living at a mad cap pace in the outer world to an inner focus, which is challenging but less hectic. One of the few experiences I have in the outer world is going out for lunch. I really appreciate this experience. It is important for me to convey this appreciation to the waiters who make this small break in the day enjoyable. I really make an effort to remember their names. I often engage them in conversation, finding out about their lives. I want them to know that they contribute to the quality of my life.

This is how I can live with appreciation every day, never losing sight of what others do to help make my life easier and more enjoyable.

Step One: Take an inventory of what you take for granted.

Begin with your closest relationship. What do you get from her/him? Look beyond the sex, companionship, taking half the workload and common interests. What about the intimacy, support, little acts of love, caring for your well-being and psychological health?

Next, use your daily reflection to look at your family members. How do you rely on them or assume that they will be there for you? Isn't there an inherit stability or sense of support, even if not expressed?

How about your key friends? In what ways do they affirm you, provide you non-judgmental acceptance and even allow you to be the imperfect person that you are? Look at how they go beyond selfish intents in valuing and caring for you.

For whatever your work is, look to see the tangible and intangible benefits you receive. A periodic paycheck is an obvious appreciation for your job performance. Other things many of us take for granted are our company benefits and a work environment which includes the tools we need to do our job. In addition, more and more often we are even given growth opportunities while being paid for our current job duties. And don't forget your boss, where it's usually easier to focus on the negative and not the positives to appreciate.

What do we take for granted about our community? Most every community is moving forward, no matter how slowly, to make life easier and more convenient for all of us. As a group, our community is dealing with issues to better our life. How often do we recognize and give thanks openly and honestly?

Finally, many people feel that their patriotism is appreciation for our larger society and government. But loyalty does not always mean appreciation. How often do you truly appreciate your way of life, the freedoms and opportunities that our society provides, as well as the protections created through both a well-equipped military and police force. Is it not similar for our fire protection, forestry service and in today's world, homeland security?

In this exercise, explore what you have in life to appreciate. How is the glass more than half-full? Let your appreciation fill the glass further. And if you're having trouble here, if the glass is less than half-full, still document what you can.

Insights: _____
_____
_____
_____
_____

_____
_____
_____
_____
_____
_____
_____
_____
_____
_____.

    In order to bring the power of appreciation to the forefront, let's use our active imagination in a key exercise.

Step Two: Visualize what life would be without each of the above.

    Take one day for each of the key people/areas described above in your life. Actively imagine seeing yourself living without each: a close relationship, family and friends, work and a larger community. Allow yourself the time to feel your reactions with each missing. Feel the void or absence.

    Most all of us can remember a time in the past or present without a close relationship. Even though there are times when we do enjoy being alone, imagine how your life would be different without a close relationship for the rest of your life. Begin with your present circumstances, what would be missing in your environment? What activities would change? What would you be trying to do to find a close relationship?

Now visualize yourself much older. See yourself having been alone for many years. What's missing? What does your closest relationship bring you that you'd rather not live without?

Another good example many of us have experienced is that of being without work. Losing and/or looking for a job is one of the most anxiety prone experiences in life. Use your imagination to visualize not being able to find a job that fits your skills and abilities. What job would you have to take?

An interesting phenomenon that has affected the attitude of young people today relates directly to finding a job. Outside of the Depression, job availability and growth has generally been good. In the last decade or so, it takes a college graduate in America an average of almost two years to find a job in her or his field of study. Many are now returning home to live with their parents after college. So, it should be fairly easy to imagine not being able to find appropriate work for an extended time.

In your visioning, see the changes you have to make to live life with any of these five key things missing. How differently would you drive your car if you knew that there would never be anyone patrolling the streets? In the western United States, what if there was no regulation of water rights? In any urban area, what if there was no regulation on pollution?

For those of you now missing a close relationship,

family, friends or work, use your reflection time in a positive way. Let your active imagination change the circumstances. Positively visualize a great relationship, good friends or a joyful and loving family. See yourself in a great job that utilizes all of your skills and abilities.

From this positive visioning, let's use a truism to learn why you don't now have what's missing. The truism is this: we each create the life we want, whether consciously or unconsciously. So, what do you need to change to fill this missing part of your life. Is it the current situation, or yourself? How can you use your I Am to better approach relationships or family? Do you need to cultivate ways to demonstrate your skills so you can find the right job? In the end, appreciate the fact that you can proactively change your life and appreciate your future success.

Insights*: (How do you better appreciate your loved ones, your fellow workers and service people in general? How do you better appreciate those who do jobs you would not do? And what about those local businesses there for your convenience?)*

Step Three: Proactively express appreciation at least six times per day.

> *"The roots of all goodness lie in the soil of appreciation for goodness."*
> *Dali Lama*

A good way to start this exercise is by focusing on what each friend or close relationship wants or values. Go beyond yourself to their needs and desires. What would make life easier or more enjoyable for them?

Regarding your partner, would she cherish more attention, affirmation or relatedness? What about an intimate evening without the expectation of sex? What about suggesting she have a girls' night out while you take care of the kids?

On the other hand, would he enjoy you spontaneously initiating sex? What about letting him host a guys' night where you prepare some goodies beforehand and then leave? Let your appreciation show you an out-of-the-box or ideal way to show your love for him.

Since we're learning to develop and express appreciation in all areas of our life, the key here is not big, overt and obvious expressions (i.e. diamond ring, two dozen roses, new golf clubs, etc.) of appreciation. At some point these may be perfect expressions. For now, make your appreciation heart felt and genuine expressions of your gratitude and love for who they are.

As to your friends, create a thoughtful way to recognize who they are. How can you help one who is struggling in life? How can you, with appreciation, be a catalyst for a positive change in their life? How can you create a positive change for them instead of a slow sympathetic support of their current circumstances?

Regarding family, how would they enjoy a contact from you, an affirmation of each as a parent, sibling, etc.? It can be as simple as a spontaneous card or phone call. It could also be just spending time with them. For your children, how can you affirm who they are? Do they need your praise for recent accomplishments or more freedom as recognition of their maturity and growth? Give some quality thought on how you can be out-of-the-box.

As you move through each day, see how life is full of ways to help make your life easier, better and more enjoyable. By making our life easier and more convenient, each of the people we meet provides us an opportunity to focus more on our spiritual practices. Appreciate them for it. A smile, a thank you or a positive comment about their service goes a long way. Learn how to express appreciate for the little things each day.

In your community, look for how you can utilize local establishments and services more. It could be your local coffee shop, grocery, retailer, restaurant or bookstore. You might have to give up one of your conveniences (e.g. internet shopping) to support a shop or service that you recognize benefits your local area.

For our larger society, there are a myriad of ways to express our appreciation. Try driving the speed limit as appreciation for our roads, traffic protections and law enforcement. Be more aware of things like pedestrians entering a crosswalk or cars trying to merge with traffic. Start obeying more laws, even if they seem silly or unneeded. Send a letter to the editor to express your appreciation for a government official, or a new law or a recent changed policy.

Insights*: (Which expressions of appreciation elicited the greatest response? What did you learn about appreciation during this exercise? How did it make you feel expressing appreciation more often?)*

_____
_____
_____
_____
_____
_____
_____
_____
_____
_____
_____
_____
_____
_____
_____
_____
_____
_____
_____.

Step Four: Create three ways to live appreciation daily and make them a habit.

It is generally understood that takes three weeks of repeated behavior to develop a habit. Over the last week or two, you have expressed at least 40 to 50 ways that you appreciate life. Choose three ways you can make them a part of your daily life.

If you need to, put a note to yourself on your refrigerator to remind and affirm your new spiritual practice. And remember, this is about your spiritualization. So don't look for or expect a particular response or credit for what you're doing for your growth.

As your new habits become automatic, feel the change in how your environment responds to you. Look for the difference it makes in the people around you. Also, become aware of how your attitude is changing for the positive. Feel how life's small disappointments are much less relevant to you now versus the past.

Insights: _____
_____
_____
_____
_____
_____
_____

Extra Credit: From *The Power of Appreciation*, reread Chapters 4 and 5 on using appreciation to transform your life. Then, utilize the five steps outlined to transform a specific area of your life. Choose an area that you had difficulty with during our spiritual practice of appreciation. Continue utilizing the five steps until you see a significant change take place in both you and the situation. Then, document your insights here.

Insights:

Trust

"Learn to trust your own judgment, learn inner independence, learn to trust that time will sort good from bad..."
                Doris Leasing

Most every religious or spiritual teaching encourages us to "trust in God." Unfortunately, the mass of humanity does not. We revere and place on a pedestal those few individuals throughout history who have demonstrated true trust in God. In most cases, humans split their trust in God with trust in the day-to-day influences from our outer material world.

Webster's Dictionary provides many definitions for trust. Let's see how this split trust is reflected in these definitions of trust. One definition is a "firm belief or confidence" in a person or thing (e.g. faith). By equating the Christ energy with Jesus, many Christians genuinely place their faith in a spiritual person.

Trust is also our "confident expectation, anticipation, or hope" (regarding the future). Except for hope, most individuals place their anticipation and expectations for the future in normal outer world sources including government, employers, community, family and money. We tend to hope that God will take care of the future, but we do not rely on God.

Another definition of trust is "to place in keeping, care, custody," or putting our "reliance" on. Don't most of us usually split this trust between giving our spiritual keeping to God and material keeping and care to ourselves? Do we not usually rely on our legal system to bring us justice; law enforcement to bring order; our military to bring us peace; our 401(k) to bring us security and our employment to bring us material rewards?

Learning to trust in a spiritual source for all aspects of our life is definitely a challenge. Like everything in spiritual growth, small continuous steps produce far more success than periodic leaps into the abyss. Also, just as the essence of God relates to us through intermediaries, we can take a significant spiritual step by learning to trust in various spiritual intermediaries.

These intermediaries can be in the form of a person, like Jesus or Buddha. It can just as easily be an aspect of Spirit including Love, Truth or Peace, for example. It could also be relying on the teachings of those role models we see who have relied on their spiritual source during their life.

The power of prayer and meditation can also create a powerful channel for us to trust in. This channel can open up a means of communication with the Holy Spirit, Christ or our spiritual Father (i.e. Brahman, Supreme Self, God Within). Unfortunately, until we have purified our Personality, we may not receive and/or understand these communications clearly.

This spiritual practice is a way to begin to transition our trust from outside of ourselves to our spiritual source within. It is most powerful when combined with prayer and meditation. Our transitionary source here is in the form of **spiritual principles**.

Spiritual principles are symbols of spiritual energies. We can use spiritual principles to transform ourselves and our life. As symbols, principles provide a fairly easy way to see Spirit working in Matter. As we become confident in relying on principles, we can remove our reliance on outer creations in matter.

Christ is an excellent example of a spiritual symbol. As communicated in the *Holy Bible*, "I am the truth, the way and the light." By simply taking each of the major sayings of Christ, we can use them as spiritual principles in which to live our life.

Frankly, many feel living Christ is either too difficult or even inappropriate for this day and age. Learning to live by Love in every situation is a powerful spiritual practice, but more difficult for some human spirits. Our spirit, or God Within can be qualified by either Love, Will, or Intelligence. Like the variety of spiritual principles, Love can be easier for some to embrace while Intelligence or Will can be for others. This is why there are many spiritual principles, some related to Love, others related to Will and some related to Active Intelligence. Choosing to utilize certain spiritual principles is a good way to align more closely to our own inner spiritual source. It does not reduce the value or importance of any spiritual principle.

In reviewing various ways to define a principle, I conclude that a principle is a fundamental truth or motivating force upon which others are based. Principles are essential types of energy upon which all others are built. They become the origin or cause of things in matter. They are the essential elements which produce specific effects.

More specifically, our objective spiritual principles originate from the Spiritual Plane, which is conditioned by Truth. This spiritual Truth is an aligning energy which catapults our orientation to a new level and reality. These spiritual principles reflect an objective synthesis (transcendence) of ideas, theories and understanding. All spiritual principles, therefore, exhibit a completeness in themselves.

So what makes a principle spiritual? A principle is spiritual because it originates from the spiritual energies within our universe. This separates them from material principles (i.e. concepts, rules or truisms) that originate within our material universe, whether mental, emotional or physical, etc. Just because a particular principle works, does not make it spiritual. So remember, material principles more closely bind us to matter. Spiritual principles free us from matter and facilitate our living by

Spirit.

Tanya's Wisdom

My upbringing was different from most in that I practically raised myself. As a result, my values were of my own creation. If you add growing up seeing individuals flirt with the wrong side of the law and trying to live just beyond the justice system, my values were colored, to say the least. If anything, I lived with a survival attitude as well as take what you want.

This attitude was slow to change and took some lessons along the way. One day standing in the checkout line at a convenience store, the clerk gave me an extra twenty by mistake. I noticed it immediately, smiled smugly at my good fortune and went on my way. Later that day after having lunch at a local diner, I realized I had lost $20.00. I was crushed, my windfall had disappeared. I thought it very strange; got 20 lost 20.

As time went on, I continued to see the connection between my actions and the resulting backlash. I began to realize that there must be something wrong with my actions. I, therefore, started to modify my attitude toward right and wrong. I began to look more closely at my actions. In the past, it had been wrong to take from individuals, but not from a corporate business. Now it was wrong to take anything that didn't fully belong to me.

I later learned that karma (what goes around, comes around) also applies to less tangible things. After getting a new computer, a friend from college came by to test drive it. As he was leaving, he showed me all the great programs he had downloaded, free. I asked if that was legal, he laughed. I decided, hey it couldn't hurt. A week later, my computer crashed. I learned a new level of what was right and the pain of what was wrong.

These lessons and many more have taught me so much about my attitudes, and my-self. I have now come to a place where I fully trust this principle of what goes around, comes around. I am not ruled by fear of what will happen, for I have found a friend that brings balance to my life.

As with our other spiritual practices, the best way to gain Wisdom here is through experience. So let's use the following exercises to learn to trust in spiritual principles.

Step One: Identify the spiritual principles you most cherish in life.

Use the next week or so to review and reflect on the enclosed list of spiritual principles. If you have been reading the *Worldwide Laws of Life*, include these principles also. Try to feel each principle provided. Highlight those that resonate with you, or lift you up in some way, energetically or mystically.

Next, review your highlighted list to determine

why each resonates with you. Does it relate to a life lesson for you? Does it represent a missed or fulfilled opportunity? Does the specific principle jump out for you because you've seen it operate for others, in their failures and/or successes in life? Does the principle touch your heart because you wish you could live it, or that you know you need to live it?

When you have finished, narrow down your list to the most powerful 3 to 5 spiritual principles for you. Be sure it's those spiritual principles that relate to you and not how others should behave. As always, this is about your growth.

My Spiritual Principles: _____
_____
_____
_____
_____
_____
_____
_____
_____
_____
_____
_____
_____
_____
_____
_____
_____
_____
_____
_____.

Step Two: Review your daily life based on these spiritual principles.

The easiest place to begin here is by observing life around you. Watch to see how these principles operate. Try to see where people or circumstances would be improved if your spiritual principles were consciously embraced.

For example, observing our boss at work can show us how his/her life, both positively and negatively, has been affected by the spiritual principle of "what goes around, comes around." Many times an individual's attitude and treatment of other employees relates directly in how they're treated by superiors regarding promotions and recognition.

It's fairly easy to see through movies and television how individuals overcome adversities to be very successful, while others, lamenting about their situation, never move forward in life. These are examples of the spiritual principle of: "it's not the cards we're dealt in life, but how we play the hand."

So, use your spiritual principles to see a new dimension in life, both positively and regressively. If you have trouble seeing these operate during the day, then use your daily reflection to review the lives of those around you. In many cases, it will be easier to see where everyone around you is not living your principles, but relying on their normal conditioning to live life.

Now reflect on your own life. Seek to identify at least one daily activity where you live each of your defined spiritual principles. Become closely aware of how these principles operate in your closest relationships and work environment. Then reflect on how your current circumstances reflect your spiritual principles, or whether your life reflects your past attitudes, orientation or behaviors. Simply, how have you been instrumental in creating your life circumstances based on living or not living your principles?

Here's an important place where we need to keep mindfulness about our spiritual principles. The energy of each principle will show you how it operates, if given a chance. So, try not to succumb to the normal daily mind chatter and busyness of outer life for this exercise.

Let the energy show you instances during the day where each principle is working out around you. Watch life, let in insights. See how much you can learn about where you put your trust. As always, try to document your insights on a daily basis.

Insights: _____
_____
_____
_____
_____
_____
_____
_____
_____

_____

_____

_____.

Step Three: Implement one new behavior for each defined spiritual principle.

With your reflections, it should be fairly easy to identify at least one way that you can better live your spiritual principles. As with our earlier spiritual practices, we have our relationships in life, our work situation, our community and the larger society to draw from. In addition, our new behavior could be related to how we perform our spiritual practices, or even just living day-to-day.

A good example is the spiritual principle: Energy follows thought. How often do we let our mind chatter drift to negative thoughts of others or current life circumstances? By stopping our daily mind chatter, we create the opportunity to direct far more energy into appreciation for others, joy in life or love of God's creation.

Here's another good place to implement these new behaviors with little fanfare. You do not need to explain what you're doing. Over time, the benefits of your new behaviors will become obvious to those around you. If you can, choose a situation where your new behavior will bring about positive change or help resolve an ongoing problem. In many cases, your change can be a catalyst for others' change. And yes, you might meet some significant resistance around you. But as long as you don't have an expectation of their behavior regarding your principles, their resistance will fade fairly rapidly.

New Behaviors: _____
_____
_____
_____
_____
_____
_____
_____
_____
_____
_____
_____
_____
_____
_____.

Extra Credit: Take each of your spiritual principles and identify two extra new behaviors which will facilitate you consciously living them daily. Set your goal to work on these until each becomes a habit. Remember, this means consciously living each daily for at least three weeks until it becomes an automatic part of your behavior.

You will find significant resistance within you as you proceed with this extra credit practice. So, put up a list on your refrigerator to remind you. Check off your list as you go along. You can even grade yourself each week on each behavior until you're consistently passing. This may take quite a while to complete, but it is a powerful way to become spiritually focused every day.

Insights: *(Which behaviors met the most resistance within you? What did you learn about how your new behaviors changed others around you?)*_____

_____
_____
_____
_____
_____
_____
_____
_____
_____
_____
_____
_____
_____
_____
_____.

Spiritual Principles

What goes around, comes around.

Give what you have, get what you need.

Energy follows thought.

It's not the cards we're dealt in life, but how we play the hand.

The whole of growth is greater than the sum of the parts.

Growth is about dealing with failure, not getting success.

Love is not the only way, only the most important. Meaning in life is a stepping stone to Spirit.

As we live in life, we live spiritually.

As we judge, we will be judged.

We must give up control to gain grace.

What we don't live in life, we will never have.

Spiritual success is about passing, an A performance means a lack of challenge.

Spiritual growth is a balance of living the process with getting to the end results.

We must give up freedom of choice to gain free will.

Truth is always relative to our reality.

There are no coincidences.

The first 20% of effort will bring us 80% of the results. Conversely, 80% of effort is required for the last 20% of growth.

Integrity supersedes honesty, truth and rightness.

The more we resist life, the more we give up God's lessons.

God always brings us what we need first, and what we want second.

Your life reflects yourself, change your life to change yourself.

Our faith defines the limits of our life.

Peace comes from dealing with conflict, not avoiding it.

This too shall pass.

Growth is about difficulty and ease, ebb and flow and giving and receiving.

Whatever we deal with within our self, we bring a gift of Spirit to our self.

Serenity is not freedom from the storm, but tranquility within the storm.

Beauty is a dramatic way of bringing meaning and Spirit into life.

What would Christ do.

Prosperity

Prosperity is our gift from God, it is for everyone. It is our choice whether to make it spiritual or material. Prosperity is what brings meaning in our life.

Ever since Charles Fillmore and Earnest Homes defined the basis of New Thought and a contemporary "new age" Christianity, becoming and living a prosperous life has been at the core. Unfortunately, today most of humanity relates to prosperity in material terms. They interpret, as *Webster's* defines, that prosperity is "good fortune, wealth and success."

Much of Christian basis for prosperity comes from the Old Testament. What's consistent about the teachings of prosperity is that we prosper by being with and keeping our covenant with the Lord. When the Lord is with us, we shall prosper.

Each of the prosperity suggested readings in the appendix provides various activities and practices to gain prosperity, whether material or spiritual. Consistently, these are practices including such things as tithing, praying, giving to others and living purposefully. What has become more confused during the 20th-century is that somehow these spiritual practices will bring material prosperity.

The truth is that a spiritual orientation will transform what prosperity is to you, especially material prosperity. Unless you have a unique life destiny, material prosperity comes from being very materially oriented, not spiritually oriented. If success to you is about material prosperity, then these books will help you gain it. They can also be used to gain spiritual prosperity, without a focus on material results.

The spiritual practice of prosperity with this guidebook is about transforming our attitude and orientation to prosperity. In this spiritual practice, we will learn how to put our focus on spiritual prosperity, while letting go of our typical orientation of prosperity in terms of security, comfort and happiness. By doing so, we will fulfill another definition of prosperity as: "to succeed, thrive and grow in a vigorous way." A key insight here is that:

> *As long as we focus on gaining material prosperity first, we will never gain spiritual prosperity.*

Tanya's Wisdom

Until I was in my late twenties, I lived below the poverty level. This isn't quite as terrible as it seems, but I did have to watch my pennies. By my mid thirties, I had married and my husband and I had gone from scraping by to a prosperous life. It was then I began to take stock of my life. I realized there was a void. It took me some time to figure out why.

As I looked around, I was living the ideal life. We had come to a place where money wasn't an issue. We had a large house with a bit of land. We both drove luxury cars. We traveled several times a year. We were financially secure!

When I looked at my relationships, I really could not find anything missing there either. I had a good relationship with my husband. I had a large group of close friends. We got together socially most every weekend. My relationship with my family was the best it had ever been. My father would housesit for us while we traveled and stayed for extended periods. My family often gathered at our house for holidays and vacations.

I was happy, safe and secure. I was the picture of prosperity. Well, except for that void I mentioned, there was something missing. The more I tried to ignore it the more obvious it became. I lacked a connection with God.

I began to make changes in my life to fill that void. Relationships that were only based on a social connection began to dissolve as I changed my focus. This freed up more time and energy for new activities and opened doors for more meaningful relationships. I spent hours reading to gain spiritual understanding. I participated in workshops and classes. I sought out individuals of like mind, spiritually.

The biggest thing I realized was that I was going to have to leave my marriage and the material world it consisted of. I could not live the type of spiritual life I needed within its confines. I had to walk away from my material world and find my spiritual life.

I have left behind that material life, but I am still the picture of prosperity. However, this picture is quite different. I don't have an endless cash flow, but my financial needs always seem to be provided for.

I don't have many friends. Those relationships have been replaced with an inner relationship with spirit that far exceeds any outer relationship.

I am now married to a man who shares my spiritual focus. Our goals run hand in hand. Our life is set up to provide the time and support for each other's spiritual practices. I am living a spiritual life that not only meets, but exceeds anything I could have imaged.

Step One: Evaluate your attitude and orientation to living a prosperous life.

As we have before, begin this exercise by looking

around you in day-to-day life. Who do you see in others as being prosperous? What makes them prosperous? What did they do in order to gain this prosperity? For individuals you know, find out specifically whether it was hard work, inheritance, luck or a combination of factors. Was it easy for them, or did they achieve prosperity with great effort and difficulty? And very important, what do they consider prosperity to be? How much of it is about material prosperity, and how much is about spiritual prosperity?

Next, where do you feel prosperous in your own life? Is this prosperity in terms of security, comfort or happiness? Or is it in terms of how you have grown, thrived and succeeded, either materially or spiritually? Besides having worked hard, what else has helped you in these areas to become prosperous?

Is there a subtle or obvious sense of pride in the status of your bank account and investment portfolio, your weekly paycheck, your college degree or recognition in the community? Is there a sense of satisfaction in your children's accomplishments, your importance within your social group, your partner's accomplishments or even the spiritual people you have met and studied with?

As you are reflecting in this exercise, try to become aware of your conditioning and/or current attitude that makes prosperity mean material wealth, success and good fortune. Instead of judging this, feel through how it pervades your outer conditioning and influences in life, whether they be family, friends, community, society, or even the spiritual community.

Insights: _____

_____
_____
_____
_____
_____
_____
_____
_____
_____
_____
_____
_____
_____
_____
_____
_____.

Step Two: What needs, both spiritually and materially, are not being met in your life.

To begin this exercise, focus on where you feel a lack of prosperity. Without looking ahead, right down what specific areas of life this is for you and what prosperity would be if fulfilled.

Unfulfilled Prosperity: _____

_____
_____
_____

_____
_____
_____
_____
_____
_____
_____
_____
_____
_____
_____
_____
_____.

As you review your list, how you do you feel about these specific areas in your life? Can you say that your needs are being met, it's just that you don't feel highly fortunate or successful? Or do you have a specific need that has not been met in each area?

For example, you may feel you do not have enough money to live prosperously. But, your income does pay your bills and provides some excess that you use for your spiritual practices and leisure activities. Is it possible that your needs are being met, but you're still dissatisfied? On the other hand, if you're living from paycheck to paycheck and feel you can't afford buying some new books or going to the next spiritual event, then not only are you not prosperous, but your needs are not being met.

To complete this exercise, use the following list as a reminder of some of the ways we can organize our orientation to prosperity. For each area, rate whether you feel highly prosperous, moderately successful or that your needs are not being met.

Material needs:

Lifestyle: *clothes, transportation, socializing, entertainment, home, travel, vacation, hobbies.*

_____
_____
_____
_____
_____
_____
_____.

Relatedness: *supportive loving family, rich partnership, caring friendships, open and honest communication outlets.*

_____
_____
_____
_____
_____
_____
_____.

Recognition: *work success, work contribution, positive feedback, fair compensation, community recognition.*

_____
_____
_____
_____
_____
_____
_____.

Security: *paying the bills, contributing to savings, creating reasonable debt.*_____

_____
_____
_____
_____
_____
_____.

Spiritual needs:

Spiritual time: *daily spiritual practices, daily reflections, daily reading and study.* ***Spiritual freedom:*** *freedom to focus on yourself, your growth and spiritual development.* _____
_____
_____
_____
_____
_____
_____.

Spiritual relatedness: *group interaction, spiritual mentoring/objectivity, supportive communication.* ____
_____
_____
_____
_____
_____
_____.

Connection to Spirit *(soul, energy, teacher, mentor, God) An open, rich and powerful relatedness to one or more.* _____

_____
_____
_____
_____
_____
_____
_____.

Vision of next step: *knowledge of where you are, where you're going and some sense of future destiny.*

_____
_____
_____
_____
_____
_____
_____.

Step Three: Evaluate each unmet need.

For each unmet need above, use your daily reflection time to see and feel your attitude and orientation today. How much time you spend each day trying to fulfill each unmet need? How do you focus on each: physical work, physical activities, emotional anxiety, daily mind chatter, wishing for the future, thoughtful planning, imaginative visioning of success, feeling that success is around the corner or knowing that it is in God's hands? There is no right answer here, only your honest reality.

For each of us, our unmet needs affect our attitude and orientation in life. Spiritual transformation is about either changing our attitude or changing our need. Prosperity is the results of either. So, how do your unmet needs affect your behaviors, the way you interact with others and the way you see your life today? Do they condition your outlook on life, and if so, how?

Finally, are there any similarities between your unmet needs, materially and spiritually? For most people, there are key similarities and differences between unmet spiritual needs and material needs. And if you're like many, your spiritual needs take a backseat to your material needs. Conditioning says that we should fulfill material needs first, then spiritual needs. If this is true for you, then it's time for a spiritual reorientation.

Complete this exercise by documenting your insights regarding your unmet needs, both spiritually and materially.

Unmet spiritual needs: _____
_____
_____
_____
_____
_____
_____
_____
_____
_____
_____
_____
_____.

Unmet material needs: _____

_____
_____
_____
_____
_____
_____
_____
_____
_____
_____
_____
_____.

Step Four: Define at least one change to meet each unfulfilled spiritual need.

Here's where we make our spiritual transformation. Realize, we normally spend at least 40 to 50 hours per week (work) focused on meeting some of our material needs. So, let's maximize the use of the rest of our time on our spiritual needs. Even more powerfully, let's reorient ourselves regarding our material needs to expand our spiritual prosperity.

So, first start with your spiritual time. What time during each week do you spend overly focused on your material needs? Where do you self-indulgently waste time each week? Good examples are talking on the phone, catching up with friends or family, watching mindless TV (not all TV is mindless), video games or window shopping (the internet is an excellent 21st century medium).

     A perfect first step is to reduce (or eliminate) watching or hearing the news. Experience validates that the slow evolutionary pace of our society can be kept up with through very little attention and focus. Take this time and simply reorient it to one your spiritual practices, like reading spiritual oriented books.

     Next, look at your relationships. It should be fairly easy to see where you invest time and energy in relationships that should be changed (or even ended). This focus can be reoriented to opening up new opportunities for spiritual relatedness.

     Even better, how can you turn a social relationship into a spiritualizing relationship? It can be as simple as focusing on supporting positive changes that each of you are making, sharing spiritual insights on what you've read or watched recently that has a positive spiritual focus. Get creative here to spiritualize a relationship, whether the individual knows it or not.

     Regarding freedom, where have you accepted undue responsibility regarding your outer world aspects in life? Is there an imbalance in your home life responsibilities? Even simpler, where have you assumed others' expectations regarding your balance between self and others? Here is a place to break free from these restrictions without neglecting key responsibilities. We all need alone time to help us bring balance in life. Get your partner or family to help you bring this balance for everyone's benefit.

Concerning money, do you have a spiritual budget as well as a material budget? If your immediate answer is, oh my god no, it's time for a change! Work with your partner or family to commit to a certain budget for spiritual practices. Remember, this can be attending movies, reading books and other activities which everyone can enjoy. And like Nike says: Just do it!

Finally, there isn't any reason why all of us cannot have some understanding on where we are and where we're going. Make a spiritual practice out of periodically identifying your next spiritual event, mentoring experience or new spiritual practice. Put some dollars in your spiritual budget and create the time to make it a part of your daily life.

If this all sounds good and fine, but very challenging, then maybe it's better to set some goals; those you can focus on in the short run and those you identify for the future. It's exciting and satisfying to be able to mark off a goal you've accomplished and move on to the next in your predefined list.

Spiritual Changes: _____
_____
_____
_____
_____
_____
_____
_____

_____
_____
_____
_____
_____
_____.

Extra Credit: This extra credit exercise is about changing your life to resolve your unmet material needs. And the key is **resolve**. Take each unmet material need and resolve it. The resolution might be giving up your orientation to it. It might be your decision to redefine what prosperity is for you. It can even be making the commitment to fulfilling it, even if it means a major change in life.

To begin with, your lifestyle is an easy change. Creative ways to deal with clothes is by buying them at second hand stores. There are some great creative ways to dress with clothes from the 20's, 30's, 60's, etc. Oh, and if you are not aware of it, buying winter clothes in February and summer clothes in August is a great way to save 40%-75% on retail.

Take your older car (many become vintage antiques) and get it detailed. Keep it clean and love it for what it gives you, transportation without debt. Also, think of entertaining "on a dime." A simple winter picnic in your house for friends can be creative and joyful (a poster of two on the walls of a summer situation goes a long way).

If you lack prosperity in relationships, give up the

ideal of the perfect partner. Define a partner that would absolutely love you on first sight. Look for that person and see the potential, rather than the limitation. Also, it is better to have one or two supportive loving friendships than a myriad of social pretenses.

Finally, redefine what a loving family is. If you do not have it, then create it with those who will best fulfill it. Remember what Christ said in *Luke 8:21*: "My mother and my brethren are these which hear the word of God,..."

Fulfilling unmet needs in the workplace is normally accomplished by changing jobs. More powerfully, it is about confronting issues with love and integrity. If you're not receiving adequate compensation or recognition, then find out why within yourself and within your circumstances. A perfect example is the work at home parent. Conditioning devalues the contribution made by this very valuable role within the family. If you're accepting this devaluation, then change your attitude and behaviors as a first step toward changing others.

When it comes to money, most individuals only relate to prosperity in terms of how much they have. Spiritual prosperity transforms money into energy and a resource; a resource for spiritual practices, spiritual progress and blessing those around you through your giving. Here you might have to make tough decisions about what expenses to eliminate to be able to pay your bills each month. Or deciding how you can create some savings, even if just a little, while pursuing how you can tithe monetarily. And finally, determining what the spiritual power of debt is, not just living a lifestyle beyond your means.

Spiritual steps toward resolving material needs:

Faith

Faith is demonstrated through the way we live life, in the little things, not in the big gestures, and not only at times of crisis.

The description of faith given above may not seem to be consistent with ways you've related to faith in the past. *Webster's Dictionary* helps us better understand this unique orientation to faith with various related keywords including: trust, confidence, reliance, security, certainty, assurance, conviction and fidelity.

It's pretty easy to relate these keywords to our orientation in day-to-day life. What do we put our trust, confidence and reliance in? Is it God, Spirit or other spiritual source or our family, government, 410K and societies' expectations of us? Faith, therefore, relates to putting our trust, confidence and security in something beyond ourselves, whether something spiritual or something mundane.

In order to define Faith, we can look to how faith has been described in past spiritual writings. From the *Holy Bible,* Christ makes repeated references to individuals "having lost their faith." This is having lost their communication or communion with God or Spirit. A second set of references by Christ are related to being healed or "made whole by faith." According to Christ, Faith is our communion with Spirit that heals us and makes us whole.

Based on this orientation to faith, Faith becomes the nurturing healing characteristic of Spirit that produces a trust, confidence and reliance in the success of Spirit's intent for our life. This spiritual definition of Faith is consistent with how Mystics have lived and taught us about Faith since before Buddha.

If faith comes in how we live day-to-day, then we are healed and made whole through our daily actions, not by some grand gesture or commitment. If we choose to live life based on our past conditioning, then our trust and reliance comes from our past experiences. If we choose to rely on the guidance and support from our spiritual source, then we will probably live contrary to normal humanity. In actuality, when we choose to live by Spirit, our new experiences confirm our faith contrary to others.

From this new orientation, Faith becomes how we are healed and made whole by God through living in communion with Spirit. A subtle, but key word here is

living. It's not just enough to have faith, we must live faith.

Unfortunately for most humans, faith becomes important in moments of crisis. Most pray to God for help when they feel they can't handle it themselves. These sporadic insignificant acts are not living in communion with Spirit. Therefore, most are not healed or made whole in resolving or living through the crisis.

Unless we change both our behaviors and our intent, faith will always seem elusive and only for emergencies. This spiritual practice is designed to help us understand our orientation to Faith and to begin the process of living faith. As Joseph Addison put it:

> *"Faith is kept alive within us, and gathers strength, more from practice than from speculation."*

In addition,

> *"The essence of Faith is fewness of words and abundance of deeds."*
>
> Baha'u'llah

Tanya's Wisdom

As I began to focus on a spiritual life, I realized it would be impossible for me to remain in my marriage. I was certain the material and social fabric of my current life would severely restrict my spiritual pursuits. With this in mind, I began to check into divorce.

I quickly faced the conditioning of my culture. The idea of divorce to pursue a spiritual life was counter to everything accepted in society. A spiritual life was found in church and had a strong basis in both a material and social life. Therefore, the idea of leaving my marriage for a spiritual life was a reach for everyone I knew, including myself.

My ex-husband was a normal southerner. As our life progressed in a male dominated society, his power in the community grew. Being the wife, mine did not. I spent 10 years of our marriage supporting his career, raising our children and building the beautiful home we lived in.

When I first talked with a psychologist, he advised me that in the traditional town where we lived my spiritual beliefs (which had become a bit new ageish) might lose me the custody of our two girls. I became terrified. This was not acceptable.

I then called several attorneys in the town we lived in. They all refused to handle my case because they knew my husband. I went on to call several in the large city near us and was mortified by what I was told. If I didn't have at least $25,000 sitting in an account in my name, I might as well forget it. I wouldn't be able to pay for attorney fees.

Everywhere I turned, I met opposition in moving forward with the divorce and I had never even mentioned it to my husband. Every day I lived with a bit of terror as I tried to set this in motion. It seemed hopeless.

Finally, I JUMPED. I couldn't take the fear any longer. I jumped off the cliff into an abyss that looked so bleak, it would be fatal. There was no choice. I knew it was the right thing to do. So I placed myself in God's hands. Finally, I said "I want a divorce." I didn't have an attorney, all I had was faith.

Then the miracle happened. In that dark abyss just beyond sight lay waiting a pillow to catch me. My husband and I had an amicable divorce. I survived. Faith in God had pulled me through.

Step One: Take time during this period to read about individuals who have lived spiritual Faith.

Think about someone you have either read about or learned about in the past that resonated faith. Obtain a book to read about this individual, or use the Internet to find some suitable articles. Good examples include Mahatma Gandhi, Mother Teresa, Martin Luther King, Joan of Arc, etc.

As you're reading, focus on how they actually lived differently from those around them. How they walked the walk, not just talked the talk. See how she/he relied on a non-material basis for life, whether love, peace, Christ or God, for example. Also try to see how each strengthened their faith as they lived out their life. See how their life was not easier, but **more difficult and**

fulfilling.

A few are born with abounding faith. For the rest of us, we are healed and made whole by our actions, especially in the face of not knowing. Giving up our reliance on money, family, society and those we love is like stepping into the abyss of which only Spirit can support.

An example in my life might provide some further insight here. My spiritual mentor once gave our group of ministers and ministers in training an assignment. We were to hitch hike across the country from the South to the West Coast with only five dollars. Setting out that day required us to act without knowing what might happen. My experiences over that 56 hour period helped me be healed and made more whole in my attitude and orientation about humans.

Individuals that helped my wife and I ranged from a truck driver to a post hippie group laughing and smoking from Arizona to California. We spent one night sleeping in the back of a station wagon parked in a gas station. We also faced the rejection of asking for a ride at dawn in a truck stop in the middle of Oklahoma. But our willingness to act without expectation paid off as one man who had turned us down, later picked my wife and I up on the road and took us across two states.

Even more significant about this is that we never said why we were doing what we were doing. It never occurred to us to try to gain support from humans through human nature. A number of the other individuals on this assignment ended up making placards which said they were "hitchhiking for the Christ." The ease of their trip demonstrated where their faith was, the sympathy and/or empathy of humanity.

Insights:_____
_____
_____
_____
_____
_____
_____
_____
_____
_____
_____
_____
_____
_____
_____.

Step Two: See the suffering in those who lack spiritual faith.

    The easiest place to begin here is with those that you know. You might even know someone who obviously walks around in pain. Beyond this, try to observe someone you know who tends to always have a negative attitude about things. Their comments always tend to take a negative turn. They can easily see how the glass is always half-empty. For this person, use your feelings to feel their pain and suffering. You'll find there is something empty inside. They will rarely talk about how they communicate or rely on an inner source.

Another good example of lack of faith relates to being the victim. For these individuals, it is always someone or something else that is responsible for their life situation. It always seems that there is nothing they can do to change things. It's as though life has it in for them.

Finally, observe and reflect on someone within your life that tends to be highly critical of others. Criticism of others is rooted in self-criticism. At the base of this self-criticism is a lack of communication with Spirit. The God of judgment is always one's own judgment. The God of love and grace has already forgiven us of our failures and weaknesses. Here again, see and feel the degree of pain, suffering and lack of faith in these individuals.

Now, for the next week or so become aware of the same suffering around you in life. In urban areas, street people are excellent examples. Within the work environment, observe those who you know dislike their job. More subtly, watch for individuals you encounter in public who exhibit a disastisfaction with their life situation. Try to become aware of how they suffer with their disastisfaction. Many people subtly live always comparing themselves to others.

Use all of your senses to become aware of how almost everyone is suffering. For some, it is a lack of self-esteem. For others, it's a lack of security or lack of love in life. Some suffer from not fulfilling the expectations of others, while some suffer from unfulfilled needs. Observe those around you until you can realize that in most cases it stems from a lack of faith. This lack of not being whole stems from a lack of communion with a higher internal spiritual source.

Insights: _____
_____
_____
_____
_____
_____
_____
_____
_____
_____
_____
_____
_____
_____.

Step Three: Examine where their faith lies.

As you direct your mindfulness (intent filled observation and reflection) toward the suffering in those you know and the rest of life, become aware of where they put their trust, confidence and reliance in life. Is it seeking after material gratification or possessions? Is it continually trying to be what others expect of them? Is it that conditioned pursuit of always seeking happiness, security and comfort? Or even a co-dependent reliance on others?

This exercise is about seeing the extent to which most humans put their faith in objective materialism and other humans. Many will even say things like "Jesus is my savior" or "God will take care of me," but continue to live day-to-day just like everyone else. The key in developing spiritual faith is building our connection and reliance on an internal support system within ourselves. Through this inner support system, we are able to "move forward, even without knowing." Our faith allows us to "relax into this vast space of *not knowing*."

Insights

Step Four: Search your heart for the changes you feel or know you need to make.

Now, let's look at our own orientation to faith. What do you rely on most? Is it like those you've observed around you? What brings you a sense of comfort and peace? Use each of your observations and insights from the previous exercises to look within yourself.

In your life, when was your faith strongest, in times of crisis or in times when things were going well? Which of these periods were you most intent and committed to your spiritual practices? When have you received your greatest healing in life? Have you been made whole through your spiritual practices or are you still struggling to be made whole by Spirit?

Using your daily meditation, go deep within yourself to identify those actions or changes you have resisted until now. Have you felt a sense that you simply cannot do it yet? Or is it that you don't feel it's possible for you to do it? Is it implementing a new spiritual practice or resolving an ongoing problem? Is it making changes in your relationships or changing your life to live your passion?

What have you avoided within yourself until now? Is it changing your own attitude about something? Is it to change a habit or behavior that inhibits your growth?

What do you see as too difficult to pursue or change about your life today?

_Insights______
_____
_____
_____
_____
_____
_____
_____
_____
_____
_____
_____
_____.

Step Five: Jump off the cliff!

During this period of self reflection, watch for a dream that can help lead you in this key step in faith. Many times, this may be a dream of fear, for fear is a major obstacle to faith. This leap of faith is about choosing one of your key insights that touches your heart the most. And then, simply take the step or steps to jump off the cliff of your own rationality into the abyss.

For help, look to see how this act, or actions fit with our spiritual principles from our trust spiritual practice. But frankly, this leap of faith will require acting without knowing in many cases. And very important here, is to resist getting support of others around you. As it was with my hitchhiking across country, it's about relying on your own inner guide. A simple way to communicate your new behavior is to say: I need to do this for me.

Naturally, you will probably meet significant resistance or opposition to your actions. If so, persevere. If there is little resistance, praise your faith.

Another key to this leap is not to expect immediate results. Your faith will show you its spiritual results over time. And remember, material results follow spiritual healing. Even though the outer changes may not seem positive to others, it's the inner changes that are meaningful. Trust that Spirit will help those involved to be healed through your act of faith.

Leap of Faith: _____
_____
_____
_____
_____
_____
_____
_____
_____
_____
_____
_____
_____
_____.

Mindful Loving

"The path to love begins in our own past and its healing, then moves outward to relationships with others."
How to be An Adult in Relationships

The above quote comes from a book which defines love as the "possibility of possibilities." From the author's (David Richo) experience, "Love is experienced differently by each of us, but for most of us five aspects of love stand out." His book and process defines a path to love which can change and heal our past and current loves. This can be a powerful step in embracing Love as a part of Spirit.

I conclude that humans experience love in three main ways. First, within birth relationships love is instinctual and assumed. The second is by what is called falling "in love." The third is by relationships in which love is allowed to grow over time. Many assume that these experiences of love are the same in each circumstance. On our path to love, we find that it is not the same love, only the expression of Divine Love at various levels within our humanness.

If this is the case, then what is Love? Various perspectives provide different definitions of love. Philosophically, "Love of God holds the universe together; the single principle permeating all things; a dynamic factor in cosmic change." Erich Fromm in *"The Art of Loving,"* defines love as "an *attitude,* an *orientation* of *character* which determines the relatedness of a person to the world as a whole,..."

Theologically, love is Charity, the "most sanctified in of all virtues." "Love alone unites us with God, delivering all unreserved to God." Divine Love is that which "draws those whom it seizes beyond themselves; a burning fire." Love is that which "spurs on that soul to union with the transcendent and Absolute Light."

One metaphysician, The Tibetan, provides a middle ground (or synthesis) by defining Love as the "free-flowing, outgoing, magnetically attractive force which leads each pilgrim home to the Father's House." I simply define Love as the coherent attractive force which makes all things whole. Within our human experience, it is abundant generosity and our capacity for relatedness.

Practically, how does this Divine Love relate to our human experience of love? In our relationships, it is either an inherent state, or a condition we move in to. Simply, it is something we are **in** that includes attraction, a willingness to give and a sense of well-being. Unfortunately, this is another example of a human experience and state which we must spiritualize.

So, how do we spiritualize this love? There are many ways to do this. Here, I choose to describe it as simply as possible. It is to transform our attitude and orientation from **in** to **out**. Like much of spiritual growth, it involves a reversal. We move from in love as a state or condition of experience to a position of acting out of Love. Out of Love, we express this coherent attractive force through our abundant generosity and selfless giving. Our Love becomes unconditional, not based on our experience.

Reread this last paragraph until you can feel the reversal involved. This dynamic change of orientation is our key to becoming a truly loving person. David Richo's five keys to mindful loving is an excellent way to begin. Obviously, our current and past relationships are a fertile field for us to learn to express out of Love. Out of Love, we give more abundantly and unconditionally. Through the attractive force of Love, our giving draws to us what we need to learn and grow.

Tanya's Wisdom

Sometimes it is easy to remain blinded by our own issues. Many times they lay dormant, waiting to rear their head. As I began to look at the five A's and evaluate how I practice them, I was surprised.

I felt that I used the first three: attention, affection

and appreciation quite well. I would even say I was in the process of a nice pat on the back when I turned my focus to the last two: allowing and acceptance. I have three children, a son who is grown and married and two younger daughters. Looking at my relationship with my son, I would say that I applied all five to this relationship. During his stumbles in life, I was there to support him, but didn't judge him. I took the philosophy that this was his journey and his soul brought what he needed. This attitude worked well for us. We grew to be great friends.

Looking at my relationship with my daughters allowing and acceptance seem much more difficult. Here is an example. As my oldest daughter became an older adolescent, our relationship began to take on a friendship, just as my relationship with my son had. I was her confessor. That is until she became involved in her first truly serious relationship. As she began to tell me things, I no longer was allowing or accepting. I was judgmental.

This surprised me. As I've looked deeper, I know a part of this is based on attachment. I wanted our relationship to remain the same. But she was moving her focus from her family to her new boyfriend.

I then realized when my son had stumbled, I knew it would work out. But with my daughter, these stumbles seemed direr. I began to wonder if I was one of those individuals who thinks that boys should be treated different from girls. Yikes. I found this very hard to accept in myself.

But there was a deeper truth in the difference

in my attitude toward her verses my son. I wanted her to have a different life than I had led. Not that my life hadn't served me, but the hard knocks school of life is quite painful. I wanted her to go to college, have a career and do all the cool single adult things I had planned to do before I had to get married at sixteen.

Her plans hadn't changed, she still planned to go to college. But my attitude had. I wanted to control the situation, to make my plan work. I now see my lack of allowing or accepting in my attitude. Some would say this is only natural, that I only wanted the best for her, which is true.

But it is her life to live, not for me to relive mine through her. I have to allow her to stumble and fall, if that is where her life leads her. I have to accept that her choices are hers to make, not mine. And you know, if I hadn't fallen so many times, I would not be where I am today. My failures have certainly served me well in getting me to make a deep spiritual commitment. I must allow spirit to help my children, just like it helped me.

Step One: Identify your closest relationships in life.

The key to close relationships is how relevant and meaningful they are in our life. It is also by the level of our commitment to the relationship. Your boss can be a close relationship because of your work commitment. A family member can be a close relationship because of its relevance to your past or present. Also, a friend can be a close relationship because of how meaningful they are in your life.

Let's organize our list according to various classifications of relationships: partner, family, work, friendship, social, community, etc. In this step, try to identify the three most important in each category (except partner), but no more. For example, within your work commitment the three might include your boss, a customer, a subordinate, a vendor or service provider, a coworker or even a fellow volunteer. A partner can obviously be your spouse or lover, or possibly a pet, or even your business partner in certain circumstances.

Closest relationships:
Partner: _____

Family: _____

Work: _____

Friends: _____

Social: _____

Community: _____

Step Two: Identify those you give the least attention to and increase your level of attention.

Once you have identified those that get the least of your attention, reflect on and evaluate which of these individuals thrive on attention. Of this group, which persons do you give the most and least attention to? Are there any specific reasons for the different levels of attention? You might find that for individuals who demand attention, you purposely do not give them the attention they demand.

This exercise is about providing more attention to the key person that thrives on attention. It is not about fulfilling wants or demands for attention. The key word here is thrive. Those who need attention, come out of themselves based on attention.

> *"Attention from others leads to self-respect. Acceptance engenders a sense of being inherently a good person."*

With greater attention, they're usually willing to give as much as they get in close relationships. When you interact with them, they're not draining of your energy, but usually help create more joy, empathy and love through the interaction. On the other hand, those that demand attention are draining and the ongoing experience is not fulfilling, either joyfully or lovingly.

Based on your evaluation, create a way to increase your level of attention to the one person who needs it most. Simply, this can be initiating having coffee once a week to catch up. It can be identifying an activity you both enjoy to periodically engage in. Exercising together or sharing a sporting activity can be a good idea.

Of course, if there is a spiritual practice or activity you both can participate in together, that's ideal. On the other hand, for some it can be as simple as an ongoing e-mail communication regarding life, books you have both read or current relationship issues. The point here is to increase your level of generosity and giving of your time to interact with them.

Your increased attention: *(What method of attention works best for he/she? How do you see them thriving on your attention? What have you learned from this change of orientation?)*

_____
_____
_____
_____
_____
_____
_____
_____
_____
_____
_____
_____
_____
_____
_____
_____.

Step Three: Based on your giving of affection in relationships, make a new habit of affectionate giving.

The easiest place to begin here is with your partner. Affection is normally an expectation and need within a partner relationship. There are many ways to increase your level of affection. But remember, this is about their needs, not your brand of affection.

For example, if you're a typical male, your orientation to affection includes sex. It might certainly address your need for affection, but not necessarily your female partner's. Therefore, resist making your affection about sex. Remove that element. Let her initiate sex, if that's her inclination. Make your focus about loving her and selflessly giving her intimacy. Touch her more often and openly in front of others, if she responds positively. In addition, increase the amount of affectionate time and focus you give her during sex. And yes, this will take significant effort and selfless orientation to fulfill.

> *"...love is not a sentimental feeling but a conscious choice to give and receive in unique and often challenging ways."*

If your female, identify what your male partner needs regarding affection. In many cases, it does relate to sex. Therefore, let your affection easily and enjoyably transition to sex. Be willing to be more creative and less inhibited in sex. Explore his body without qualification or judgment, if he responds to it. Learn what kinds of affection he thrives on and give it to him. Also, in many cases, open and sincere affection in front of your kids or others may be just what he needs.

Beyond your romantic partner, define which of your other intimate relationships would best respond to additional affection. For that person, increase your affection to them unsolicited. Identify and implement a new habit of giving them affection. This can range from back or foot rubs to hugs and kisses, from touching affectionately to just being couch potatoes in front of the TV more often. Of course, the key here is to provide additional love in the form of affection to the person who will most respond to this form of love.

Insights: *(What noticeable changes in self-esteem, loving communication, sense of well-being and overall attitude have you observed in each person from your increased affection?)*_____
_____
_____
_____
_____
_____
_____
_____
_____
_____
_____
_____
_____.

Step Four: Increase your expressions of appreciation to the individual who needs more appreciation in life.

It should be fairly easy to identify this person based on your spiritual practice of appreciation. In many cases, it will be the person who most lacks self-esteem or self-reliance. Reflect on what you really appreciate about this person, whether it's their qualities, behaviors, accomplishments in life, their uniqueness or even their orientation toward your relationship. Then begin periodic expressions of your appreciation.

Of course, these expressions can be verbal affirmations of your positive view of him/her. Additionally, identify what you can do to help them with their self-esteem or self-reliance. This could be shared activities to help support them. It could be ideas for new behaviors or changes that will bring an increased sense of well being. It can be as simple as heartfelt expressions of love while you're with them.

The important factors here are your sincerity and follow through. Individuals dealing with issues of self-esteem or self-reliance can smell insincerity. You can also compound the issue if your expressions stop after a few weeks. It's also good to remember that it may take a significant amount of time for you to recognize changes. If little changes over time, then communicate empathetically about their difficulty in expressing their positive qualities.

Your expressions of appreciation may need to take the form of coaching. A coach is someone who provides a high level of support and direct involvement in helping someone to learn and become proficient. In the case of self-esteem or self-reliance issues, the person needs to learn to appreciate oneself more and act accordingly.

Appreciative affirmations: *(What are the positive qualities, uniqueness and accomplishments this person has demonstrated in life? How can they use these to further enhance their life? How can I participate as a loving and intimate friend?)* _____

_____
_____
_____
_____
_____
_____
_____
_____
_____
_____
_____
_____
_____
_____
_____
_____.

Step Five: Lovingly help the individual who can most benefit from your allowing.

There are many types of non-allowing in life. In a majority of partner relationships, a common form becomes one's unfulfilled expectations of the partner. One partner tends to want to control the other's behavior to meet their self-imposed expectations. Another common occurrence in many relationships revolves around judgmentalness. It's easy for many to have opinions about what someone else should be doing better or right, while simultaneously not taking steps to resolve their own issues.

Psychologically in many romantic relationships, each partner unconsciously wants the other to live out their contra-sexual self. This means the husband wants his wife to live out his feminine nature, while the wife wants him to live out her masculine nature. Many times this translates into the wife expressing most of the emotions and intimacy in the relationship, while the husband must focus on providing the stability, support and identity for the wife. Overtime, many of these couples go through a process of realizing that their spouse has never really understood who they really are.

Expectations is a form of non-allowing that pervades many family and work relationships. We are subtly or obviously coerced into living out someone else's expectations, rather than lovingly getting freedom and independence. Lifestyle issues become psychological battlegrounds, like what or how much one should eat, drink, or self-indulge in. Non-allowing can also revolve around one's love-life (the right or wrong boyfriend or girlfriend) or health issues including smoking, drinking or use of drugs. Is it really our place to expect someone else to live our lifestyle, even if it is healthier?

Finally, most everyone suffers from non-allowing in society. Each generation tries to mold the next generation to their idea of who they should be and how they should behave. The renegade, rebel or non-conformist rarely meets with freedom and independence.

Who suffers most from non-allowing? Review your close relationships and choose the one that stands out. Then decide how you can allow them either to be more of who they are or to live with more freedom or independence. Without going to the extreme of supporting very unhealthy behaviors, decide how you can remove your inclination to want to control them to either to fulfill your expectations or judgments about them.

An obvious example relates to children. It is a rare occurrence that a parent is willing to provide loving supportive freedom to an adolescent who doesn't meet their expectations. Seemingly good intentions turn into issues of control and non-allowing. It has been demonstrated that most values are instilled in children before the age of seven. There is little chance of success in trying to instill our values into a teenager who is naturally developing a sense of identity through challenges to authority.

In the case of a teenager, creating more relatedness through shared activities while removing your conditioned judgment and expectations will create a more positive loving atmosphere in your relationship. In many cases, you'll see the rebelliousness diminish while more open communication flourishes. And even better, when painful growth events occur in their life, they will more often turn to you for love and advice. Further, this new orientation is not about removing all limits.

> *"Limit-setting makes it safe for you to be yourself. Paradoxically, we can't achieve freedom without limits. They are the holding environment in which we flourish."*

Beyond this example of a family offspring, the same applies to most all your close relationships. Creating an atmosphere of freedom and independence with loving support for the other person will naturally increase the love between you. You will also naturally see more open intimate communication. As it always is with love, your generous giving will always attract love in response.

Insights: *(Why does this individual suffer from non-allowing? How have you contributed to it through your attitudes and behaviors? What can you proactively do to provide added freedom and independence without judgment or expectations?)* _____
_____
_____
_____
_____
_____

_____
_____
_____
_____
_____
_____.

Step Six: Change your attitude and interaction with the person who will best benefit from increased acceptance.

Acceptance is similar to allowing, but more focused on who someone is in their current reality. It's not about allowing behaviors or vices, but lovingly accepting one's total character, both positive and negative. As David Richo further puts it:

> *"Acceptance means we are received respectfully with all our feelings, choices, and personal traits and supported through them."*
>
> *"In acceptance, (we) are embraced as worthy, not compared to (others) but trusted, empowered, understood, and fully approved of as (we) are in our uniqueness."*

In our society, there are far too many cases where this acceptance has not been given to children, females and siblings. This lack of loving affects many for their entire life. It tends to only compound the inertia involved in making positive spiritual changes in life. When we are conditioned with a lack of love from others, it's even more difficult to embrace love from God.

As we have before, review and reflect on your list to choose that individual who needs more of your acceptance. Here's an opportunity to provide unconditional love through support and affirmation of who this person is with or without their behaviors. Realizing that none of us is perfect, but only express degrees of imperfection, lovingly embrace this person with empathy and compassion. The degree to which you can do this will measure your degree of acceptance.

Let this loving acceptance guide you in defining how you can express your love better in this relationship. You will usually find that this individual will respond fairly quickly to your change of orientation. As love flows out, love flows in.

As your acceptance grows, the ease by which you accept everything in life will also grow. You'll be amazed by how you can become a center of love expressing out to a world which will absorb it like a sponge. And as a sponge receives water easily, the world can also give love easily.

Insights: *(What feelings and new perspective of love have you gained from your unconditional acceptance of this person? How do you find it easier to live this unconditional love to a greater degree?)* _____
_____
_____

_____
_____
_____
_____
_____
_____
_____
_____.

*With the total of these five keys to mindful loving, how has your capacity to give and receive love increased?)*_____

_____
_____
_____
_____
_____
_____
_____
_____
_____
_____
_____
_____
_____
_____
_____
_____
_____.

Job's Journey

*Every branch in me that beareth
not fruit he taketh away: every
branch that beareth fruit,
he purgeth it, that it may bring
forth more fruit.*

John 15:2

      Much of our conditioning from childhood includes the concepts of good and evil, right and wrong. Obeying the law is good, going to church is good, honesty and loyalty are good. On the other hand, disobeying your elders is wrong; greed, lust, envy and vengefulness are evil. Drawing clear lines helps us as children to behave and grow positively in a world that does not live by just good and right.

      On the Spiritual Path, we need to internalize our own values and learn to live them with integrity. This is an important spiritual process which separates us from normal human nature and, therefore, brings clarity of what is good and right. In this process, evil takes on a whole new meaning. One good definition of evil becomes: "that which should be subdued and controlled, but which is allowed to act." Wrong and evil become personal terms that do not apply in the same way to everyone.

Embracing our quote from *The Gospel of John* can be greatly helped by understanding the story of Job. *The Book of Job* attempts to wrestle with and resolve "the age-old question: Why do righteous men suffer, if God is a God of love and mercy?" Understanding Job's journey will also help each of us persevere as God continues to purge us along our journey in becoming a living spirit.

It is easy to accept that there must be a period of transformation and transmutation of our lower human nature while on our spiritual journey. The simple concept is that this is how we become good and righteous. Unfortunately, until we truly penetrate and internalize Spirit, what is good and righteous evolves, just as our Soul and Personality grow to become individualized expressions of Spirit.

This spiritual practice is designed to help each of us realize that our growth need not just come from what happens in life, but also from our attitude toward our experiences. The spiritual reality is that we will all become Job to one degree or another. Giving up our simplistic conditioned ideal regarding good and evil and right and wrong will highly benefit us as Spirit purges our fruit along the way.

Tanya's Wisdom

As I considered where my life experiences were similar to Job's, I drew a complete blank. I couldn't see a correlation at all. So, I turned to Jef to see if he had any insights and he did.

After marrying Jef, it became clear that we needed to move to Colorado. This was a difficult choice, but I had known for years that moving west was eminent in my life, even before meeting Jef. I had hoped it would be much later after my kids were grown, but life doesn't always flow the way we would like.

It became my goal to find a way to take my life in the South with me to Colorado. I tried to condition my daughters to make the move with me. I looked into ways to continue my MFA so I would have a secure future, even-though I knew my future wasn't teaching art, but helping others pursue their spiritual potential. I tried to arrange everything so my life would be the same. The only real change would be my location.

One by one, all my efforts fell apart. The tighter I held onto my life, the more it slipped away. I realized that my purpose for pursuing my MFA was purely for security and it would be very close to impossible to continue with it once we made the move.

My daughters decided they preferred to stay in the South where they were comfortable and everything was familiar. They decided to live with their father. I was devastated. Since I was 16, my main role in life had been mothering. To have that torn from me was heart wrenching. Some might say I had a choice to stay, but

I knew I didn't. I knew that to remain in my current life would be a spiritual failure. And I might not get this opportunity again.

So I left the South behind me. I left my children, my family, my friends, my source of financial security and moved into an unknown future with nothing of past familiarity. I took with me an emptiness that seemed unfillable.

Now, I have come to realize that although I don't see my children as often as I would like, my relationship with my daughters is closer than ever. We truly are friends. I know that they are better off having the opportunity for a regular life as they grow up. As far as the rest I left behind, I know that this is my spiritual path and what God wants me to do in fulfilling my destiny.

Step One: Re-read *The Book of Job* with a new perspective.

The Book of Job spends an inordinate amount of focus on debates between Job and his friends. Job's perseverance of his innocence and integrity does contribute to understanding Job's journey. The greater power and importance of this book comes in the first and last few chapters. So, you need not get hung up in the extensive details along the way.

Having said this, reading *The Book of Job* with a new perspective means not focusing on the suffering, but to focus on the growth. What did Job need to learn during his experiences? Assuming that God is loving and merciful, what was God bringing Job for his benefit?

If this seems like a callous orientation, it may be time to realize that as the Buddha taught: all of our existence is suffering. Unfortunately, most of the time we choose to suffer. At certain places in spiritual growth, we suffer in order to grow.

As you move through this book, become aware of what growth Job received. Part of it relates to understanding those around him. Part of it relates to maintaining his integrity in the face of adversity. Part of it relates to his attitude and orientation. A part even relates to him becoming conscious of his weaknesses.

Part of our new orientation to Job is our view of Satan. If good and evil are much more abstract, personal and obtuse than we are conditioned to believe, let's change our orientation to Satan for this exercise. As you read *The Book of Job*, think of Satan as a part of God that helps us grow in Spirit through adversity.

Insights: *(What qualities of Job needed pruning? What did he need to realize about his family and friends? How did Job's relationship with God change?)* _____

_____
_____
_____
_____
_____

_____
_____
_____
_____
_____
_____
_____
_____.

Step Two: When and where have you suffered in life?

Lets begin this by making a list. Either start with your childhood and move forward or your last few years and move backward. Continue adding to your list until you have reviewed your entire life and identified at least 29 experiences of suffering, or enduring adversity. If you find it easy to identify many more than 29, then narrow your list to the 29 most intense experiences of pain or adversity. If you literally can't define 29 after significant review, then better appreciate your life.

Next, review your list and identify what ended these experiences. To summarize your review, identify general categories of these changes including, but not limited to: outside influences, your change of attitude, a choice or decision that was made, a spiritual step you took, life changing circumstances, a death or a specific person's influence.

Suffering: _____
Change:_____

Step Three: Evaluate your growth in these experiences.

With your above list, identify which of these experiences you feel brought you growth and resolution, and which did not. Using all of your senses including your feelings and intuiting, define what made some of them growth experiences, rather than just suffering. Key differences in growth versus non-growth can be your attitude, orientation or a steps in growth including:

Taking a step forward   Self responsibility
Your change of attitude   A spiritual influence
Making a personal sacrifice   A personal choice
Playing the hand you're dealt An act of faith
Giving up being the victim

As you look at your list, see the trends in what produced growth and resolution of your suffering. Then look at those experiences which you identified as non-growth and see if you can see trends in how your suffering ended.

Insights: *(List the trends that produced growth and ended your suffering. How did you inhibit growth and extend your suffering through your own attitude and actions in your non-growth experiences?)*_____
_____
_____

_____
_____
_____
_____
_____
_____
_____
_____
_____
_____
_____.

Step Four: Take proactive steps to resolve at least 2 ways in which you currently suffer.

If you're like most, past experiences of suffering which you see as non-growth are probably continuing or incubating today in some form or fashion. Just enduring adversity usually creates fear for the future. This fear will usually bring more adversity, just as Job's greatest fear became his reality.

Here is a place to use your past growth experiences to help you resolve your current suffering. What worked in the past to resolve your pain or adversity can more easily work today. It will take a growth attitude. It will be more easily accomplished if you can move forward with confidence in God's love and mercy in your life. And it will bring you a greater spiritual prosperity, just as it did for Job.

Current Suffering & Resolution: _____

_____
_____
_____
_____
_____
_____
_____
_____
_____
_____
_____
_____.

Extra Credit: Review your life today and identify all the ways in which you are suffering or dealing with adversity. Define how you can end these sufferings from your insights in this spiritual practice. Then set in motion the changes in your attitude, orientation and behaviors that will end each. If need be, set some goals in dealing with them over time. Then review your life periodically to make sure you complete your list.

Additional Current Suffering & Resolution:

_____
_____
_____
_____
_____
_____
_____
_____
_____
_____.

Changing Altitude through Changing Attitudes

"... bringing your life into balance and maintaining your spiritual equilibrium require a focused awareness and daily retreat from the stresses of the world."
   Susan L. Taylor
     Lessons in Living

   On the Spiritual Path, ultimately the only thing we can control is our attitude and orientation. By changing our attitudes, we bring our most significant changes in spiritual altitude. Similarly, the more we try to control our lives and future, the less we allow our God

Within to open up vistas of Spirit to us.

Each of us is naturally conditioned to focus on controlling our outer life. This usually becomes our natural attitude and orientation to everything. Bringing spiritual equilibrium is a very important step on the Path. To do so, we must change our orientation. But what is the attitude and orientation that we must create? The end result will finally be a continual orientation to living Spirit. That's an easy thing to say, but an exceedingly difficult thing to do.

As with much in spiritual growth, there are transitionary steps we can take. When one step becomes internalized and automatic, then we are ready for another. Unfortunately, most all of us must simultaneously live in the outer world. So, a good first step is to become unburdened by it; to allow the outer world to become our helper in changing who we are and how we live. And how to we do that? By taking to heart what the philosopher William James said:

> "The greatest discovery of my generation is that human beings, by changing the inner attitudes of their minds, can change the outer aspects of their lives."

A key step in changing our attitudes involves engaging various spiritual Forces. Giving up our natural sense of control is like closing one door which allows another to open. One door that can open brings the Forces of purpose, focus and balance into our life. Utilizing these forces for growth is one method of changing our spiritual altitude through a change in attitude.

By changing our inner attitude to one of purpose, focus and balance, we bring to bear spiritual forces which will accelerate our spiritual journey. Purpose and focus embody the 4th Unity and 5th Spiritual Planes within our universe. Balance, on the other hand is a subjective Force very close to our physicality. Each can be used and internalized as a dynamic spiritual growth step.

I conclude that to become spiritually focused based on purpose necessitates first bringing balance within our lives. The I Ching of eastern mysticism is a philosophy of "balance in life." The key to spiritual growth based on the I Ching is internally and externally balancing yin and yang. Balancing takes place at many levels in order to achieve perfect equilibrium.

One way to embrace balance is through two definitions of balance from *Webster's Dictionary*. They include: "to compare as to relative importance" and "to bring into a state of equilibrium." As you will see, in this spiritual practice we will do just that with our outer life and inner orientation.

Once we have gained a balance of spiritual versus material and inner versus outer, we can then move on to utilizing both purpose and focus to further accelerate our spiritualization. The Force of purpose will bring us into alignment with our Soul's plan. The Force of focus will bring us an actively aware center of attention and capacity for expressing our purpose from our Soul and God Within.

Tanya's Wisdom

Years ago, I was talking with a friend. She told me a particularly horrid story she had read in the news. I was surprised by my interest in the story. I turned my attention to what drew me to the news and realized I had a pattern. I noticed that my mind chatter focused on news stories. I would spend a lot of time turning on the drama in the news, whether it was some political intrigue or a domestic squabble being reported. I also realized my energy level changed as I became engrossed in the dramas. I soon concluded that I had to remove these influences from my life. I stopped taking in news in any form.

This helped tremendously. And in the 20 years since I made this decision, the world seems no worse off from my cutting out the news. It seems to be running its course without my focus. And I feel better by not being inundated by the negative energy that comes through so much of the outer world events.

With less outer world focus, I felt a growing urge to meditate. I found this very difficult. I would sit and my mind chatter would be running all over the place. I would get bored and think of a thousand things I needed to be doing. All of them seemed more important than my meditating. I tried focusing on a candle flame. I tried to get friends to meditate with me; I guess misery really does love company.

I finally came to a place where I knew I needed to force myself into a corner and really work on quieting my mind. So I signed up for Vipassana, a form of meditation. I packed my bags and went to do a "sit" for ten days. That is funny since I usually did good to sit for ten minutes. And here I planned to sit for 12 to 14 hours a day for 10 days.

There were two sides to those ten days. One, it was hell! I planned and plotted my escape a lot of the time. I knew I could carry all my bags for miles. That would be better than having to endure the anguish of sitting another minute.

On the other hand, I did get results. I recognized a shift in my energy level. I began to gain control over my mind. I didn't win the battle, but it was the beginning of a new way of handling my unruly mind. The length of time I could go without a thought increased, because I stuck with the program. This was another step forward in moving from an outer focus to an inner focus of refection.

Let's focus in this spiritual practice on the first third, that of balancing, in changing our altitude through changing our attitude. Also for this spiritual practice, you will begin Steps One and Two at the same time.

Step One: Turn off the world.

For the next week, do everything possible to remove yourself from the external world you normally live in. Begin by turning off all the ways you obtain the news. Second, watching TV at times during the week is acceptable, but at a minimum mute the commercials. And by the way, current satellite and VCR recording options create a fantastic way to eliminate the harassment from commercials and the outer objective material world it brings to us.

This week, act as though you're on vacation and going to be away. For those who will support you this week, the truth is fine. For those in question, lie with integrity to Spirit. Socially, this means skipping a week of parties or other engagements. This also includes both close friends and family. Hopefully they will respect the truth and your need for privacy this week.

If possible, take time away from your work as much as you can. If you work in the business world, take some mental health days if they are included within your sick leave. Obviously, using vacation time is an option. If you must work, then minimize your interaction with others. Take your breaks by yourself and go for a walk. Choose a week for this exercise where things will be fairly quiet. Of course, go to lunch by yourself and don't choose a week where you must work overtime.

There is nothing wrong with exercising, but use either ear plugs or earphones with suitable music to minimize the outer noise around you. Instead of eating out, order in this week, or just prepare your own meals.

Finally, if you were on vacation, you would reduce how much you do for others. Do that for this week. For those of you with a spouse and/or children, this week is about minimum interaction. So, get them to support you while you tend to hide out in your bedroom. This of course means that you will need to solicit great help from your spouse. You will probably find that based on your success, he/she will want you to reciprocate.

Step two: Enjoy the solitude!

This week is about getting to know more about your inner self. It is also about experiencing the benefits of a more balanced life. This is an excellent time for increased meditation and self reflection. If you're like most, you've got some spiritually oriented books to read. Rent some movies which either resonate with your inner self or can help you in your balancing. Some good examples include: Regarding Henry, Seven Years in Tibet, A Good Year, Castaway, Jonathan Living Seagull, The Razor's Edge (the remake) and Legends of the Fall, to name a few.

This is a good week to reflect on what brings you joy and peace, especially internally. What have you been missing or lacking that this internal experience fulfills? Embrace the aloneness without loneliness. Listen to the silence. After a period of adjustment, hopefully you'll find that the silence communicates a lot from Spirit.

As you become deeper and deeper within your inner life and can let go of the fears associated with your own aloneness, reflect on how the outer world has helped you avoid yourself. Embrace yourself, as though you're meeting yourself in the first time. This inner part of yourself can be your best friend. Make his friend your closest friend. You have nothing to lose, and part of yourself to gain.

Insights: *(Describe the part of you that you better got to know this week? What Ah ha's did you experience? What fears did you face?)* _____
_____
_____
_____
_____
_____
_____
_____
_____
_____
_____
_____.

Step Three: Define your inner attitude to your outer life.

As part of your last day, or even the next week, reflect on the inner life you could lead and the outer life that you do lead. Most of us do not have the opportunity to live every day as we choose, but for this exercise assume there are no limitations.

Begin with your inner life as you lived it this last week. What parts of it would you embrace every day if given a chance? Beyond the activities, what energy states (tranquility, serenity, peace, joy, etc.) did you value the most? How did time change during this inner experience? What would you like to experience over and over again?

Next, reflect on the way you are in the outer world from this inner perspective. What parts of you are lost in the shuffle of everyday life? How do the influences from the outer world lessen your ability to feel balanced? What would you easily give up about your outer life to live your inner life more often?

How do you see your relationships differently? What expectations of others are far less important or valuable to you now? On the other hand, what parts of your relationships take on more importance based on your aloneness this last week? What other parts of your outer life do you still, or even more cherish?

Insights: *(Besides answering the questions above, what seems less important in your life based on this inner experience? And what is more important about living the rest of your life?)* _____

_____
_____
_____
_____
_____

_____
_____
_____
_____
_____
_____
_____
_____
_____
_____.

Step Four: Define your ideal way of living a balanced life, both inner and outer.

Active imagination is a powerful tool in creating our future and communicating with our Soul. Over the next week or so, use your daily reflection time to envision and imaginatively live out your ideal balanced inner and outer life. See yourself living a scheduled daily life that embraces both inner activities with outer responsibilities. What parts of your existing life would remain the same? What part of your ideal life do you already live?

Your Ideal Vision: _____
_____
_____
_____
_____
_____
_____

_____
_____
_____
_____
_____.

Now, with this ideal vision, evaluate what parts of your life you will need to change. What parts of your outer life create the stress and anxiety which inhibits living your inner self? What parts of your daily life today restricts those key energy states you experienced during your retreat? Who around you inhibits this balanced life?

What changes need to be made to bring this ideal balance into reality? If your work must change, how could it possibly fit within your ideal? Who will be affected and how, both positively and negatively? What new habits will be required for you to develop? And even more importantly, what habits must be broken?

Without evaluating the ease or difficulty of these changes, document those changes which would be required to live your balanced life.

Key Changes: _____
_____
_____
_____
_____
_____
_____
_____
_____
_____

_____
_____
_____
_____.

Step five: Implement three changes which will bring you a more balanced inner and outer life.

As you review your list, apply the 80/20 rule. Which will bring you the most balance? Which of these changes will require the least amount of effort? Which will have the least impact on and be the easiest to gain support from others?   What new habits will be the easiest and most balancing to develop?

With these three changes, think through and define how you can gain support from your partner, family, close friends, work associates and others. Assume that you will need to negotiate in a give-and-take win-win manner to accomplish each change. Anticipate what would be a win-win for others. Anticipate that the outer world will be like a demanding child that needs your compassion and integrity in order to cooperate.

Realize, these changes will require significant efforts by you. Creating or breaking habits is always a major part of spiritual growth. We always change our spiritual altitude by changing our attitudes. Living more spiritual further requires changing our behaviors and influences from others.

Changes to Implement, When to Accomplish:

_____
_____
_____
_____
_____
_____
_____
_____ .

Extra Credit: With your vision of a balanced life and the changes necessary to live it, define at least one goal per year to make the remaining changes a reality. At least once a year, take a long weekend or up to a week to repeat this spiritual practice until you can reasonably say: I have achieved a spiritual equilibrium of living a balanced inner and outer life.

Compassionately Helping Others

"Spiritual energy brings compassion into the real world. With compassion, we see benevolently our own human condition and the condition of our fellow beings. We drop prejudice. We withhold judgment."
Christina Baldwin

    As we have been learning in these spiritual practices, our spiritual transformation comes from a new spirit oriented attitude toward many things we've already experienced and practiced humanly. Two good examples are joy and trust. The same is true for how we help others. Our early conditioning and learning provides us the basis for helping others as a human. On our spiritual Path, we need to spiritualize the way we help others.

In much of the spiritual literature we read and individuals we admire, love and selfless giving are common factors in helping others. A key ingredient is to be able to identify with the pain and suffering of others in order to help them. Our emotional connection helps us to give love while becoming a loving being. Two forms of emotional loving are sympathy and empathy. Another term, compassion, is used interchangeably as a form of loving.

It is not very difficult for us to realize that love takes many forms at many different levels within Spirit and Matter. What becomes difficult is the ability to love spiritually while transforming the various ways we love humanly. This involves moving from loving selfishly to loving selflessly. Most of the time this also involves moving from attachment to non-attachment with others' pain and suffering.

One way to make this transition is through three of the ways that love expresses: sympathy, empathy and compassion. In order to do this, we need more specific and delineating definitions of these energies. Let's begin with sympathy. *Webster's* provides various definitions including: "sameness of feeling; affinity between persons or of one person for another; a mutual liking or understanding arising from sameness of feeling" and "pity felt for another's trouble, suffering, etc.."

Sympathy is identified as a feeling or emotional state. It is a connection or relatedness based on the same emotion. Simply put, "I feel your pain and I suffer with you." Within the realm of emotionality, sympathy usually expresses as a sharing of human emotional states like sadness, grief, in-love and wanting, to name a few. Even though many use the term sympathy regarding spiritual energies, we will soon see a better term to label spiritual love.

A second emotional energy, seemingly similar to sympathy, is empathy. *Webster's* does provide a delineating definition: "the projection of one's own personality into the personality of another in order to understand the person better; ability to share in another's emotions, thoughts, or feelings." The key here to delineate empathy from sympathy are the words "to understand" and "ability."

With empathy, we learn to understand another person's emotional state better, not necessarily share their emotions. With empathy, we have the ability to share in someone's feelings based on understanding their pain and suffering. To actually share their feelings and emotional states is being sympathetic. Based on this, empathy is a higher form of love and relatedness than sympathy. Accordingly: "I can feel your pain. I can identify with your pain. I understand your pain." For myself, I conclude that empathy is our highest form of emotional love.

Now we come to compassion and its relationship to empathy and sympathy.

> "A truly compassionate heart is *not* emotional."
> Alice A. Bailey

Through Buddha's enlightenment and thorough spiritual transformation, he became a compassionate being that has helped teach humanity how to detach from the pain and suffering in the world. All matter is the state of suffering. Our journey is to dis-identify with matter and achieve a state of Nirvana as a spiritualized being. Mahatma Gandhi helps us here to understand compassion:

"Our motivations are colored by our unfulfilled inner needs, which lead to a personal investment in things' turning out a certain way. The wise, on the other hand, are free from such concerns. They do not laugh or cry at the ups and downs of the world, but maintain an inner equanimity in the face of loss or gain."

In the spiritual state of Nirvana, we experience what Alice Bailey identifies as "Universal Love which has no relation to sentiment or to the affectional reaction but is, predominantly, in the nature of an identification with all beings." More simply, compassion is a form of spiritual love characterized by objectivity, detachment and the identification with the spiritual beingness of all of us.

The purpose of compassion is to help us resolve our suffering, not live in it. Therefore, through compassion our loving supportive help to others is more like "tough love." It is an active love from our Soul which directs our Personality to provide help, or detachment, based on the needs of the person involved and the nature of their suffering.

Tanya's Wisdom

In recent years, I've begun to look at how I live my relationships. My mode of helping others became an area of concern. An example of this is my relationship with a good friend. Our way of helping each other was through "mutual support." This helping support was so colored with sympathy and empathy, we really only managed to support each other's dysfunction. We were great at this. The sad thing is we really thought we were helping. If I screwed up, I could count on her to tell me that I was doing the right thing. She could always count on me for the same thing.

This was especially evident with our romantic relationships. We both were there for each other supporting all the co-dependent, dysfunctional relationships; telling each other exactly what we wanted to hear. Several years earlier, she had supported me in a relationship for six months while I dated a man trying to leave his fiancée. So when she began dating a man who had been living with a woman for 20 years, I returned the favor and supported her for a period of time.

Finally, I came to a place where I stopped. I realized I wasn't helping her by providing a false hope about this relationship. This type of support hadn't helped her deal with past relationships. So when she asked my opinion about her current relationship, I told the truth. I didn't support her illusion.

It surpised me when she walked away from our relationship. I have never heard a word from her since. I understand I had changed the rules. I was no longer willing to help her as I had in the past and I don't think she wanted that type of friendship.

I had another friend whose friendship went

back to our mid-twenties. We were the best of friends. We had seen each other through those party years, relationships and all the sundry things we stumble through life dealing with. We each began to explore our spirituality and our relationship shifted to this new focus. This became the core of our friendship for years.

There came a time when she walked off the spiritual path. I talked with her many times concerning this. I tried to help her find her way back to the spiritual focus she had left behind. It was to no avail. She could see where she had gotten off track, but was unwilling to take the tough steps to get back on track. I knew that remaining in this relationship would be a form of support for her failure to move forward. Finally, I stepped aside. I did so with love and compassion.

Letting go of the relationship was to allow her to make her choices and to deal with the consequences of those choices. It was a very difficult thing to do. I still think she might get a wake-up call that will bring her back to her path. And if she does, I'll be there to support her on her spiritual journey.

Step One: For all your key relationships, review how you have, or can help them move forward positively.

Another way to put this is to review your

relationships (partner, family, friends, work, community) to see how they reflect either one or more of sympathy, empathy or compassion. Through your daily reflection, become aware of how you condition the relationship in moving forward compassionately or supporting them to remain the same sympathetically or empathetically. What are yours' and theirs' expectations in the relationship?

Look at how you communicate sympathetically with some and empathetically or compassionately with others. Identify the ways in which you help them resolve their suffering, or just live easier with their suffering. Based on your interaction with each of them, which of these three forms of love tends to dominate?

From a perspective of compassion, what would be a better place for those relationships characterized by sympathy and empathy? How can you help them with their current issues or lifestyle that produces their suffering? Do they want help, or are they comfortable where they are? Will your help be appreciated or resisted?

Insights:
Partner/Closest: _____
_____
_____
_____
_____
_____

Family members: _____
_____
_____

_____
_____
_____
_____
_____
_____
_____

Friends: _____
_____
_____
_____
_____
_____
_____
_____
_____

Work relationships: _____
_____
_____
_____
_____
_____
_____
_____

Community: _____
_____
_____
_____
_____
_____
_____.

Step Two: For those denying or avoiding your help, identify how you can become more compassionate.

Begin this exercise by identifying where you have given up helping them, or if you have. Who or what is the key issue? Has your help been based on your same feelings like theirs (sympathy), or that you understand their suffering (empathy) and would like to help them get past it? Do you continue to try to help them even though they ignore your help?

For each of these individuals, identify their key positive traits or character. Then change your attitude and behavior to communicate your support of these positive character traits. Beyond that, stop providing your help, whether through communication or behaviors. Provide nothing else but support of who they are as positive human beings.

Also, see how you can remove any criticism or judgment that you have about them. Remember the spiritual principle: "Judge not, and ye shall not be judged."

Then watch to see where their commitment to grow or change evolves over time with your new relationship with them. If it does, then provide compassionate help. If it does not, provide compassionate support of who they remain to be.

Insights: *(For each person you identify as denying your help, document three key positive personality traits or parts of their character)* _____

_____
_____
_____
_____
_____
_____
_____
_____
_____
_____
_____
_____
_____.

Step Three: For those who are resisting your help, become more compassionate to resolve their resistance.

There can be a number of reasons why they resist your help, some related to you and others not. In many cases, you have already identified those reasons that do not relate to you. They are afraid of what the future will bring if they make changes; they have become comfortable with the status quo; the changes seem too difficult to make or they cannot deal with others' reactions to the changes they would make, etc.

A key part of their resistance could also be your attitude and orientation in helping them. If part of the way you interact with them is based on sympathy, then they accurately feel that you're trying to get them to do something that you are not, or will not do. Their perception inhibits them from moving forward. They have the added weight of your sympathetic emotions to deal with.

If you have come to place of empathy for one or more of these individuals, their resistance may seem baffling. As mentioned, empathy is a high form of emotional love. Certain states of mind could be in direct conflict with your empathy. Condemnation, criticism or judgment is a good example. You could be empathizing with them emotionally while simultaneously criticizing or judging them subtly. Remember, our subconscious is a powerful thing and everyone can feel it, even if they or we do not identify it as such.

In these cases of empathy without obvious sympathy, our mental attitude may be generated from an internal criticism or judgment of ourselves. We therefore try to resolve our own emotional issue through our attitude toward others. If we can authoritatively help others, it will then be easier to help ourselves, or so the illusion goes.

If this does not prove to be the situation, then try to become aware of how other parts of you may be in conflict with your empathy. Here's a good place to get feedback from your mentor, facilitator or sounding board. If your reflection and evaluation produces no results, then the issue is probably with the individual who is resisting you.

Compassion is not a mental state, it is a spiritual form of love. Becoming more compassionate is about transforming our mental attitudes and behaviors which typically express condemnation, criticism and judgment. Being able to identify with all beings and the inherent nature of suffering in our world is a good step beyond our mental limitations. An even better one is to realize and live that we would rather not be judged by God, either in this life or after it. Do we not truly desire God's love and compassion for us?

Insights:_____
_____
_____
_____
_____
_____
_____
_____
_____
_____
_____
_____
_____
_____.

Step Four: For those who are dependent on your help, use compassion to help them resolve their suffering or dependency.

Obviously if you have small children, they are dependent upon you. Many marriage and partner relationships also exhibit a dependency of one upon the other, or both as co-dependency. Co-dependency is about living with sympathy. It does not involve compassion. If you are in this kind of relationship, then Step One should already be helping you become more compassionate.

Dependency can involve all three forms of love, especially involving children. One key effect of dependency is that it tends to drain your energy. Almost every parent will confirm how tiring it is to raise children; and not just day-to-day, but over long periods of time. This draining of energy and creating of inertia is compounded by how much we sympathize and empathize with our children.

Beyond partners and children, many relationships among friends, family, work and community involve dependency. A key compassionate solution was taught by Christ to his disciples. The spiritual principle is instead of providing fish to others, teach them how to fish themselves.

Begin reviewing these relationships by determining to what degree is each individual capable of becoming a fisherman. What areas of their dependence on you can be changed into each becoming independently self-reliant. How can you help them become independent through loving supportive change?

Here's where compassion becomes very important. Those who rely on you are not going to give up this dependency easily. They are going to resist you teaching them, both obviously and subtly. The status quo is more comfortable, less fearful and supports their desire for a sympathetic interactive relationship. Your more objective detached love is like weaning a baby off the bottle or teaching toilet training. But we all know the benefits of these types of learning processes.

Within yourself, you will also need to deal with the particular part of you that will resist this change of orientation. What do you gain in this dependent relationship? Is it emotional satisfaction, pride, ego or even intimacy and support of your own emotional weaknesses? Does your help for them improve yours or their quality of life? Both individuals in every co-dependent relationship gain something. Be sure to figure out what you're gaining with the status quo.

So, the key here is to create a win-win change for both of you. Helping them to become more self-reliant and independent is definitely a win. You'll win by the positive benefits that every inter-independent relationship brings. But be sure to identify other activities or areas of life where you can continue to reap the gains you've gotten by this dependent relationship. Otherwise, you will struggle to maintain your new inter-independent relationship. Of course, you can always move forward and transform the parts of you that needed this dependency.

Insights: : (*How does the individual's dependency on you provide you benefits? What ways of being compassionate have helped them move forward? What have you gained also?*) _____

_____
_____
_____
_____
_____
_____
_____
_____
_____.

Extra Credit: In your key relationships, take further steps to move to a compassionate inter-independent relationship with each. For those who respond to your compassion, identify where extra help is needed and give it. Identify where less help will be valuable or beneficial and then make your internal adjustment.

Adjust your orientation to their commitment and orientation. Let go of your requirements and expectations regarding your help. Give without attachment. Finally, stop giving where there are no results. Implement the spiritual instruction: cast no pearl before swine.

Decision-making

As our trust and faith in Spirit develops, our decisions obey the new direction of our life.

As part of the human species, our decision-making evolves as we mature. As infants, we are incapable of making decisions for ourselves. An important part of our learning and growth process is learning how to make choices for ourselves and deal with the consequences. Just as important a part of our spiritual growth process is the increasing opportunity to make decisions in line with our Soul's plan and destiny.

Spiritualizing our decision-making is a vital part of how we externalize our spiritual individuality. Like everything else in spiritual growth, we need to transform the various ways we make decisions based on our Personality influences.

For those remaining primitive humans on our planet, decision-making is fairly easy. It all stems from survival, relatedness and fulfilling of desires. For the rest of us, there are a myriad of influences which we must work through for every important decision in life.

An early part of spiritualizing our decision-making is about becoming aware of the driving influences in our decisions. This spiritual practice will be about gaining this awareness and focusing our decisions on those influences that assist our spiritual growth.

To help us become consciously aware of the driving influences in our decisions, let's begin by outlining a model by which we approach life as we grow. As mentioned above, primitive humans and most children are innately driven by their instinctual nature. Even though much of our growing Personality includes redeveloping instinctual parts of our mentality, imagination and emotionality, we do have an internal instinctual nature that has been very beneficial to us in the past and the present. This instinctual nature is identified psychologically as our id. Astrologically, it is represented by the moon.

This core instinctual part of our-selves includes various ways we have survived and evolved as humans including fear, the drive to survive and procreate, our herd instinct and our capacity for relatedness. Basically, these parts of us have been tremendously valuable in navigating the difficulties and dangers of living within our physical universe.

Much of our decision-making as children (which remains an influence within us as mature adults) comes from these instinctual drives. The important point to remember is that what served us well as evolving humans becomes an obstacle in our spiritual growth.

Our next step up as humans is our emotional nature. As powerful and almost as instinctual as our core primitive nature. For many, this conglomeration of emotions, desires and empathetic love drives decisions from early childhood until death. From our lowest lusts, greeds, envies and sympathies to our highest empathies, human joys and elations, many of our decisions are colored, for better or for worse.

An important tool in spiritualizing ourselves and our decisions comes from our imaginative nature. This higher energy level above our emotionality involves our instinctual fantasies and hopes and wishes, but also includes our creativity and psychic senses. Until we become conscious of our Soul's Plan, our imaginative nature provides a close connection to our Soul's unconscious influence.

The highest part of our Personality is our mentality. We are conditioned early on to make decisions by the shoulds and should nots taught to us. In addition, we are also taught to use our rationality as the driving force in our decisions. The key thing to remember here is that this is not a spiritual influence, only a higher part of our developing Personality and society.

Spiritualizing our decision-making begins beyond our Personality. It comes from our intuitive nature. And unfortunately, this influence is subtle at best until we develop our intuitive senses. In the early part of our spiritual transformation, intuitive insights are the main way we get spiritual input objectively. This is why meditation and daily reflection are so important. Subjectively, our dream life is also a powerful source of insights from our Soul that we can utilize.

The good news is that as we develop spiritually, other higher influences begin to take over. These include realizations and revelations from our inspirational nature, as well as the power of gnosis (knowing) from our higher Self. And we are not done yet. Beyond this, we even have the spiritual faculties of comprehension, synthesis and discernment to aid us in climbing the most arduous parts of our mountain to spiritual fulfillment.

As our opening quote reveals, the simplest and direct route in transforming our decision-making from Personality based to spiritual based comes in learning to live by trusting our spiritual principles and living by faith. With these two earlier spiritual practices under our belt, we are ready to transform our decision-making to align with our spiritual commitment.

Tanya's Wisdom

For a large part of my life, I have made decisions on the fly. I never really gave much thought to my course of action. Looking back on my choices, I can easily see several motivating factors, most of which were unconscious.

I bought houses, got married, and made every major decision with the same thought process, as if I were buying a candy bar. No I probably gave more thought to the candy bars.

I once bought a car far above my means because I was running with a rich crowd and wanted to fit in. Eventually I had to sell it because I couldn't make the payments. In this case, I ended up having to buy an old beat-up Volkswagen Rabbit, which was more aligned with my pocket book, but not my ego.

I bought a huge hundred-year-old house with little thought of upkeep and maintenance. I realize now that unconsciously I saw this as a way of adding stability to my life, which is something that had always been missing.

I decided to have additional children because it was what my husband wanted, even though it wasn't a real desire of mine. After two, I agreed to have a third about the same time as I began to focus spiritually. After giving it a lot of thought, I realized this was his wish not mine. And one that I would have to live with for the rest of my life. I couldn't go through with it. I realized I had been missing the point.

I now look at the whys behind the choices I make. I give a lot of thought to each decision and decide if this choice is based on spirit. It is not uncommon for my first instinct to be based on my humanness and that I need to revise my choice to align with spirit.

Step One: Identify what you consider to be your most important decisions throughout your life.

Over the next week, use your daily reflection time and/or meditation to do a mini life review. A good way to do this is to work backwards from the present. For example, if you're 35 then use each day to review five years. The reason for this is you will probably find it difficult to remember much from the first few years of your life. By working backwards, you will engage your unconscious and subconscious to help you recall your childhood years.

Write down all the decisions that were significant in your life. Include those decisions that have influenced your spiritual journey, and also those which have affected your personality life. When you have completed your review, go back to your list and rank each decision in terms of importance to your life at the time. A simple way to do this is to use a ranking of hi, medium and low.

Once completed, then select your top 11 decisions in order of importance for the rest of this practice.

11 Decisions: _____
_____
_____
_____
_____
_____

_____
_____
_____
_____
_____
_____.

Step Two: Evaluate each of these decisions and their affect on your life.

An easy way to do this is to select two per day to review during your meditation time. Why was this decision important in your life? See the positive outcomes that resulted in your life. See the stress and conflicts involved in each. See the pain and learning which was also involved.

In addition, identify who or what were key influences in your decision-making. Was it your parents? Was it your friends or peers? Has it become your partner or closest relationship? Have you had a mentor along the way who was influential? Was the person involved in your decision a key factor to your decision? Was it their influence or just their behaviors which drove your decision?

Now, use the various levels of our Personality described earlier to identify what types of energy or psychological state were key to each of your decisions.

For simplicity, the chart below can be a good reference. Was it a sense of fear or survival that drove your decision? Were you in-love or trying to fulfill some desire? Did you change your life to fulfill some hope or wish? Was it an unconscious influence that you sensed was the right thing to do? Was it a conclusion you reasoned out? Finally, can you identify a spiritual insight or intuition that motivated your decision?

Based on your evaluation, which of these decisions would you change today if you had to do it over again? Which of these decisions, if any, are you still uncomfortable with in your life? Or more precisely, how would you change the timing and/or execution of your decision? The point of this detailed evaluation is for you to immerse yourself in these important parts of your life to gain some new insights about how you have approached decision-making.

Energy level	**Motivator**
Instinctual:	fear, survival, anger, sex, herd instinct, relatedness
Emotional:	lust, greed, envy, sadness, grief, sympathy, in-love, desire, joy, passion, elation
Imaginative:	fantasies, hopes and wishes, psychic intuition, dreams of future, values, right versus wrong
Mental:	pride, vanity, logical, reasonable, should or should

not, ego, self-esteem, self-affirming

Spiritual: intuition, insight, vision, telepathic communication, realization, revelation, knowing, divine intervention

Insights:_____
_____
_____
_____
_____
_____
_____
_____
_____
_____
_____
_____.

Step Three: Identify and affirm those positive key influences which produced your best decisions in life.

With your completed evaluation, it should be fairly easy to identify positive influences and trends in your past decision-making. These key influences can be individuals, groups, community or society influences, or one or more of the motivators described above.

Define how you can affirm these influences in a positive way. See how you can increase their influence in your life. Of course, for those that are spiritually oriented, here is a good place to affirm and increase your trust in them for the future. To complete this step, identify and implement three ways in which you can increase the power of these positive influences.

Three Ways: _____
_____
_____
_____
_____
_____
_____
_____
_____
_____
_____
_____.

Step Four: For those decisions not seen as spiritually progressive, make changes to transform their influences.

In many cases, you will have at least one decision that you would do differently today or that you still feel uncomfortable with. So, focus on what you can do now to change your attitude, orientation and/or behaviors that still stem from any lingering influences.

One example might be your decision to get married. If the marriage was not a good one and it ended in divorce, this decision could be one that you are still uncomfortable with. In many cases of early marriages, the herd instinct is a driving factor. Everyone else is getting engaged, it's time for me. This guy will be a good provider and I'm so in love. In this example, you can't go back and not get married. But you can decide to stop trying to keep up with the Joneses, and stop letting infatuation influence your relationship decisions.

Of the number of motivators described above, which ones have influenced your regretful decisions? For most of us, Personality motivators become subconscious in their influence as we mature. We think we're making a decision for one reason, but there are other factors involved. "I think this is a reasonable and logical job decision." Subconsciously: "I need the security of this paycheck and am afraid I will not find a better job if I wait."

The key to uncovering our subconscious motivators is by looking at the circumstances and situations we put ourselves in. This can also include the influence of others, the basis for our relationships, or even the on-going responsibilities we maintain.

So, look at what you can change now in a positive spiritually motivated way. Look beyond the surface to see what more instinctual or personality based motivators still influence you. Then, make the decision to free yourself from these limitations and increase your decision-making based on your own spiritual nature.

Insights: _____
_____
_____
_____
_____
_____
_____
_____
_____
_____
_____.

Extra Credit: For most of us, there is a decision that we have hesitated to make within our life, but still lingers. Usually, it is a difficult step that doesn't have a clear path which will satisfy our Personality needs. Often, it is just below the surface and tends to jump in and out of our conscious awareness from time to time.

Focus your most deep meditation to go within your heart and see what decision still lingers on the horizon. Then, use your meditation and dreams to focus on this decision until you get a spiritual insight regarding it. When you do, don't hesitate to move forward. Utilize what you have determined as positive spiritual influences in your past decision-making for input on this decision. Also, bounce your impending decision off of your mentor, facilitator or sounding board for objective input.

Then, cast your fate and faith to following your spiritual directive.

Filled with the Spirit

*If ye continue in my word,
then ye are my disciples indeed;
And ye shall know the truth
and the truth shall set you free.
          John 15:26*

   One of the most difficult things I find for many to understand is the Holy Spirit, or Word of God. In trying to find a good book to include in our suggested readings, I could only find those that speak from an orthodox Christian perspective, in which the Holy Spirit is a he. The holy Trinity is the Father, Son and Holy Spirit, all males. It is perfectly acceptable to relate to the Holy Spirit as a person. It is just as real to relate to the Word of God as a voice, energy or communication from the mind of God.

   There are many levels of Spirit. Some parts of

Spirit *dance* with and within Matter. Other parts are yet to be manifested in a distant future we can hardly conceive of. For this last spiritual practice in our program for spiritual transformation, getting acquainted with and inviting the Holy Spirit into our heart could not be of more help. With the Holy Spirit, or "Comforter," we have a communication channel that will lead us into all Truth, whether that truth relates to us or to our outer life.

As Christ was preparing to leave his disciples, he conveyed that he had to leave in order that the Comforter could come to them. It was not time yet for the Christ energy to fully manifest within humanity. There would be a second Coming which would bring Christ's victory on Earth. Until then, the Holy Spirit would be our guide. Christ is "the truth, the way and the light." The Holy Spirit is our guide on that way into the light with truth, regardless of how we label it.

What's fantastic about this part of Spirit is that it comes to where we are, no matter how far we've come in our spiritual journey. Even with all our remaining imperfections, we can still make this part of Spirit a true friend and confidant. It doesn't care what we call it, only that we embrace it with our heart. Once we have, it will become like a beacon showing the way ahead on our spiritual Path.

There on many different ways to approach the Holy Spirit. The steps provided here have become a popular means within the Christian community for people of all levels of spiritual growth. To remain harmonious with various Christian Churches, I choose to define the term Christian as: anyone who follows the teachings of Christ. It really is irrelevant what particular interpretation (or dogma) one is most comfortable with. Therefore, to help here, embrace those teachings that speak to you and let the rest falloff of you like water off a duck's back. God's creativity will always provide a myriad of ways for us to come home.

Tanya's Wisdom

When Jef proposed the title of this chapter as Holy Spirit, I was adamant he had to change it. He laughed at my conviction in this matter. I clearly have a bias to the term. It brings to mind the fundamental churches that pervade the rural South. And you can imagine my reaction when he said the first step is to visit a charismatic church.

Then I remembered my friend Kathy. She and I were best friends in high school. As we both began to look for spirituality in our lives, she choose charismatic churches. I was confused by this, my cool friend going to one of those churches. She said she liked the spirit in the churches. I asked about all the dogmatic preaching. She laughed and said she didn't pay any attention to it, but loved the singing and feeling she got while there.

When I first met Jef, he told me he had gone to a Pentecostal church for years. What? Here is a man who believes there are many paths to spirit. And that wasn't my understanding of fundamental churches. All I could make of this was here is a man who will do anything for spirit.

As for myself, I've had spirit filled moments. Music does seem to be a very strong catalyst. One song in particular is George Harrison's My Lord. It moves me! I have several others, they inspire me to clap my hands and yell hallelujah.

I know this level of spirit hasn't anchored within me. So I guess, I might just have to check out one of those churches.

For this spiritual practice, begin implementing Steps One through Four during the first week or so.

Step One: Select and visit a charismatic church for four weeks.

The charismatic movement is the fastest growing segment within Christianity and has now become a part of many Christian denominations, even the Catholic Church. Embracing the Spirit is also a part of many non-Christian churches and spiritual groups, whether western or eastern oriented.

For this spiritual practice, I'm recommending that you find a service which focuses on the Holy Spirit. Simply, it will take you far less time and effort to fully immerse yourself in the Spirit. If you find a non-Christian service which is very spirit oriented, try it if it speaks to you. As already mentioned, you're not in any service in this practice for the dogma, but to commune with the Spirit.

Try to find a service that is highly regarded for its charismatic focus. Approach your visits with a sense of adventure. Don't be shy, get involved. Sing with enthusiasm. See yourself as a part of the group, not a separate visitor. Meet the people like you could become their best friend. Tell them you're looking for a home where the Spirit thrives!

Feel the energy as it moves throughout the service and the congregation. You're here to experience, not to evaluate. Feel and watch as the Spirit builds in the service.

If they call you forward to testify, join in. Remember, this is an adventure for you in Spirit. Express your heart, your feelings, yourself. See and feel your self surrendering to the energy of the Holy Spirit. Let your resistance flow out of you. Let yourself go.

Insights: _____
_____
_____
_____
_____
_____
_____
_____
_____
_____
_____
_____
_____
_____.

Step Two: Embrace uplifting music.

For this spiritual exercise, change the kind of music that you normally listen to. While driving, find a Christian station to listen to. Buy or borrow at least one CD of classic Christian spiritual songs; the ones that include Amazing Grace, How Great Thou Art, etc. Find other music to embrace the Spirit. Van Morrison is a contemporary artist with some powerful songs. A greatest hits can also work. Beyond the words, feel the power of the Holy Spirit being sung to the rooftops.

Some contemporary Christian music also embraces the Holy Spirit. Unfortunately, much today is more about entertainment or affirming certain beliefs than truly expressing the Holy Spirit. Again, if the singing gets into the Spirit, the words are far less important.

Insights: _____
_____
_____
_____
_____
_____
_____
_____
_____
_____.

Step Three: Read and embrace *The Gospel of John.*

      I recommend you use a red letter Bible which emphasizes the specific sayings of Christ. Read one chapter each day, both in the morning and evening. It is better if you do this right before and after sleep, or at least before and after you begin your normal daily activities.

      Use your daily reflection time to feel and embrace the messages in John, especially the sayings of Christ. Feel as you read. Use all of your senses. Let the message penetrate your heart. Christ is love and compassion, let it open your heart.

Insights: _____
_____
_____
_____
_____
_____
_____
_____
_____
_____
_____
_____
_____

_____
_____
_____
_____
_____
_____
_____
_____
_____.

Step Four: Pray each day to the Holy Spirit for guidance.

Talk from your heart as though the Holy Spirit is a person. Invite it to become a part of your life. Ask it to show you what you need in your dreams. Ask it to lead you into it, to bring you home to its comfort and peace.

Another way to do this praying is while you're exercising. Use the following mantra for the next month during your exercise program:

Step in Spirit, align to Love.
Step in Spirit, align to Love.
Step in Spirit, align to Love.
Step in Spirit, step in Spirit.

If you don't exercise regularly, start doing a 30 minute walk three or four times a week. And remember, the words are important, but the power comes in the energy and conscious focus by which you execute the mantra. You'll know if you're successful if your mind chatter ceases and you can feel the energy as you go.

Insights: *(What specific answers to your prayers did you receive? How did the Spirit show you its presence and its guidance?)*_____
_____
_____
_____
_____
_____
_____
_____
_____
_____
_____
_____
_____
_____
_____.

Step Five: Define one on-going activity to continue to embrace the Holy Spirit in your life.

If you've been earnest in this spiritual practice, there have been some powerful moments and experiences in the Spirit. For many, they experience what is called the "visitation" of the Holy Spirit. Normally, this is a very emotional experience.

Prior to the dissent of the Holy Spirit, it visits us to help us bow and surrender our personality to its spiritualizing energy. And when we truly open our heart, it floods in like a fire hose. And with this, the energy doesn't leave. It's as though we have created this lifeline of energy, which we can call on at any time, any place and in any situation. With the dissent and residency of the Holy Spirit, a major channel to and from Spirit is established within us.

Choose one of the activities that has moved you the most. Then make a habit. If reading *The Gospel of John* moves you the most, then continue your exercise with Mark, Luke and Revelations. Then let the Spirit move you in how to proceed next.

My new habit: _____
_____
_____
_____
_____ .

Extra Credit: Continue one or more of the above activities which move you in the Spirit until you have the "knowing" that the Holy Spirit is anchored and resident within you to stay.

If you return to your normal church services after this spiritual practice, introduce one aspect of the Holy Spirit into your home church services. You could sing in the choir, help with the music, lead a Holy Spirit Sunday school class. Move forward with however the Spirit leads you to help bring it to others.

Holy Spirit activities: _____
_____
_____
_____
_____
_____
_____
_____
_____
_____
_____
_____.

Epilogue

Hopefully, you began this spiritualizing program on a high note of Joy and finished it on a high note in Filled with the Spirit. During the journey, you probably had some intense moments. With the energy provided, you most likely also experienced many powerfully revealing insights into your own nature. In addition, because this is your workbook for spiritual transformation, you also felt guided without being forced and supported while being challenged.

For those of you who embraced this method of spiritual growth and mentoring, our *Living Spirit* community provides an ideal environment to continue your spiritual journey. Within *Living Spirit*, we have developed seven spiritual Paths based on both the initiatory process for spiritual growth and the astrological cycles of life. These Paths utilize the same structure as this Guidebook, step-by-step spiritual practices within Phases of growth.

Different than many spiritual Paths, these are fully Aquarian* (focused on self-facilitation and self-responsibility) with periodic inputs suggested from your spiritual mentor. They encompass the wide variety of avenues in which we pursue our life integrated with the various spiritual teachings throughout history. These Aquarian Paths for spiritual growth include pursuing your spiritual fulfillment through Living Christ, Eastern Mysticism, psychological Individuation, Astrology, Relationships, your Vocation/Career or through don Juan Matus's Warrior's Path (as documented by Carlos Castaneda & others).

For spiritual practitioners/ministers, these various spiritual Paths provide the same opportunity for facilitating study groups as does this Guidebook. Whether one pursues spiritual growth as a more traditional student of a teacher/guru/savior, or through these Aquarian Paths, outside object mentoring is always a positive way to accelerate spiritual growth. This is why the *Living Spirit* community is just as focused on helping practitioners, ministers and groups provide meaningful ways to build their local spiritual community.

As it is with God's creativity, there are many ways to pursue living Spirit. And they will all lead us home to the Father's house of Spirit.

* Aquarian refers to the energetic influence from Spirit that cycles through various Ages. The Piscean Age of the world savior lasted from the time of Buddha to the rise of science in the early 1600s. The current Aquarian Age is just beginning its Gemini phase of communication and education of Aquarian spiritual growth through self initiation and self responsibility.

Suggested Reading

Joy

> *Wake up... Live the Life You Love in Spirit*
> Compiled by Stephen E. & Lee Beard: Little Seed Publishing. 1-933063-02-5

Being I Am

> *Toward a Psychology of Being*
> Abram H. Maslow: D. Van Nostrand Company. 978- 0442038052

Appreciation

> *The Power of Appreciation*
> Noelle C. Nelson, Ph.D. & Jeanne Lemare Calaba, Psy.D.: Beyond Words Publishing. 1-58270-104-0

Trust (for reference)

> *Worldwide Laws of Life*
> John Marks Templeton: Templeton Foundation Press. 1-890151-15-7

Prosperity (any of the 3)

> *Prosperity*
> Charles Fillmore: Unity House. 978-0871593122

> *The Four Spiritual Laws of Prosperity*
> Edwene Gaines: Rodale Books 978-1594861956

The Dynamic Laws of Prosperity
Catherine Ponder: DeVorss Publications.
0-87516-551-6

Faith

Faith
Sharon Salzberg: Riverhead Books.
1-57322-228-3

Mindful Loving

How to Be an Adult in Relationships
David Richo: Shambhala Publications.
1-57062-812-2

Changing Altitude through Changing Attitudes

Waking Up in Time
Peter Russel: Origin Press.
1-57983-002-1

Compassionately Helping Others

Compassion: The Ultimate Flowering of Love
Osho: St. Martin's Griffin.
978-0-312-36568-4

Filled with the Spirit

Experiencing the Spirit
Robert Heidler: Renew Books
0-8307-2361-7

Living Spirit

Living Spirit for us is living a focused spiritual life through practical day-to-day activities embracing a wide variety of spiritual growth venues.

Our Community

We embrace and welcome individuals, groups and practitioners seeking to better express spirit within their lives. Our community is here to provide tools, techniques, programs, products and interaction to assist all in taking spirit into their everyday life.

We include, and are not limited to, those demonstrating their spirituality through eastern or western mysticism, psychology, astrology, science, education and the arts, as well as those pursuing spirituality through their business career or relationships.

Come learn how our community can help you. For

Individuals: Programs for accelerating your growth

Practitioners: Tools to better promote your practice

Ministers/Groups/Retailers: A toolbox to help build your local community

Join our community at livingspiritcommunity.net

It's **Free!**

Book Photography

The vast majority of the photography in this book comes from the collection of Jenny McCarthy. You can purchase full colored signed prints from her at thequakingaspengallery.com.

Jenny McCarthy

Jenny began developing her passion for photography at 11 when her father gave her a fully manual camera and said that she had a unique vision of things. By the time she graduated from high school, she was ready to begin her professional career in photography.

Becoming an understudy to Freddie Jones, Jenny refined her artistic talent over the next 5 years. Within a few years, she also became recognized throughout the Vail valley, Colorado for her quality photographs and artistry, culminating in becoming the official photographer for the promotion of the new Jack Nicklaus golf course.

Today, after 17 years as a professional photographer, Jenny owns and operates her own gallery in the Pine River Valley in southwest Colorado. Over the last few years she has been recognized in various national award programs for a number of her photos.

Notes

Notes

Notes

Notes